Chemical Oscillations, Waves, and Turbulence

Chemical Oscillations, Waves, and Turbulence

Y. Kuramoto
Kyoto University

Dover Publications, Inc.
Mineola, New York

Bibliographical Note

This Dover edition, first published in 2003, is a slightly corrected republication of the work originally published by Springer-Verlag, Berlin, Heidelberg, New York, and Tokyo, in 1984.

Library of Congress Cataloging-in-Publication Data

Kuramoto, Yoshiki.
Chemical oscillations, waves, and turbulence / Y. Kuramoto.
p. cm.
Originally published: Berlin ; New York : Springer, 1984.
Includes bibliographical references and index.
ISBN-13: 978-0-486-42881-9
ISBN-10: 0-486-42881-8
1. System theory. 2. Dynamics. 3. Self-organizing systems. 4. Oscillating chemical reactions. I. Title.

Q295.K87 2003
003—dc21

2003046194

Manufactured in the United States by Courier Corporation
42881805 2013
www.doverpublications.com

Preface

This book is intended to provide a few asymptotic methods which can be applied to the dynamics of self-oscillating fields of the reaction-diffusion type and of some related systems. Such systems, forming cooperative fields of a large number of interacting similar subunits, are considered as typical synergetic systems. Because each local subunit itself represents an active dynamical system functioning only in far-from-equilibrium situations, the entire system is capable of showing a variety of curious pattern formations and turbulencelike behaviors quite unfamiliar in thermodynamic cooperative fields. I personally believe that the nonlinear dynamics, deterministic or statistical, of fields composed of similar active (i.e., non-equilibrium) elements will form an extremely attractive branch of physics in the near future.

For the study of non-equilibrium cooperative systems, some theoretical guiding principle would be highly desirable. In this connection, this book pushes forward a particular physical viewpoint based on the *slaving principle*. The discovery of this principle in non-equilibrium phase transitions, especially in lasers, was due to Hermann Haken. The great utility of this concept will again be demonstrated in this book for the fields of coupled nonlinear oscillators.

The topics I have selected strongly reflect my personal interest and experiences, so that this book should not be read as a standard textbook. Nevertheless, the spirit by which the present theory is guided may stimulate those students in various fields of science who are fascinated at all by the curiosity of the self-organization in nature.

I am particularly grateful to Professor H. Haken who initially suggested that I write a book on this subject. I wish to thank Dr. H. Lotsch of Springer-Verlag for his patience in waiting for my never-ending manuscript. I am also indebted to Mrs. K. Honda for painstaking typing assistance.

Kyoto, February 1984 *Yoshiki Kuramoto*

Preface

Contents

Chemical Oscillations, Waves, and Turbulence

1. Introduction

Mathematically, a reaction-diffusion system is obtained by adding some diffusion terms to a set of ordinary differential equations which are first order in time. The reaction-diffusion model is literally an appropriate model for studying the dynamics of chemically reacting and diffusing systems. Actually, the scope of this model is much wider. For instance, in the field of biology, the propagation of the action potential in nerves and nervelike tissues is known to obey this type of equation, and some mathematical ecologists employ reaction-diffusion models for explaining various ecological patterns observed in nature. In some thermodynamic phase transitions, too, the evolution of the local order parameter is governed by reaction-diffusion-type equations if we ignore the fluctuating forces.

One important feature of reaction-diffusion fields, not shared by fluid dynamical systems as another representative class of nonlinear fields, is worth mentioning. This is the fact that the total system can be viewed as an assembly of a large number of identical local systems which are coupled (i.e., diffusion-coupled) to each other. Here the local systems are defined as those obeying the diffusionless part of the equations. Take for instance a chemical solution of some oscillating reaction, the best known of which would be the Belousov-Zhabotinsky reaction (Tyson, 1976). Let a small element of the solution be isolated in some way from the bulk medium. Then, it is clear that in this small part a limit cycle oscillation persists. Thus, the total system may be imagined as forming a diffusion-coupled field of similar limit cycle oscillators.

We now turn to a Navier-Stokes fluid for comparison. The flow may be oscillatory as in, e.g., the Taylor vortex flow (DiPrima and Swinney, 1981). In this case, however, it is apparently impossible to imagine such local dynamical units as persistent oscillatory motion even after isolation. After all, every term on the right-hand side of the Navier-Stokes equation represents "interaction" because it involves a spatial gradient. In this respect, reaction-diffusion systems bear some resemblance to thermodynamic cooperative fields which are also composed of similar subunits such as atoms, molecules, and magnetic spins or, in a coarse-grained picture, semi-macroscopic local order parameters. Furthermore, it may happen in reaction-diffusion systems that the global dynamical behavior is predicted, at least qualitatively, on the basis of the known nature of each local dynamical system, whereas such a synthetic view hardly seems to apply to fluid systems. It is also expected that the bulk behavior is much less sensitive to the boundary effects in reaction-diffusion systems than in fluid systems. This implies that in the former case the study of infinitely large systems would be of primary importance; although different system geometries and boundary conditions may provide a variety of intriguing mathematical problems, they seem to be of secondary importance at least from the physical point of view which we take in this book.

We noted the analogy between reaction-diffusion systems and thermodynamic cooperative systems. However, the former differ essentially from the latter in that each local subsystem can operate in far-from-equilibrium situations so that it may already represent a very active functional unit. It is this difference that makes reaction-diffusion media capable of exhibiting the wealth of self-organization phenomena including turbulence never met in equilibrium or near-equilibrium cooperative systems. In this book, we will concentrate on the fields of *oscillatory* units which are coupled through diffusion or some other interactions. For a variety of other aspects of reaction-diffusion systems, one may refer to Fife's book (1979a).

It may now be asked what sort of self-organization phenomena are expected in this type of field. In considering this problem, the importance of the concept "synchronization" or "entrainment" cannot be emphasized enough. This simply means that multiple periodic processes with different natural frequencies come to acquire a common frequency as a result of their mutual or one-sided influence. In some literature, the former term is used in the more restrictive sense of the oscillators' phases also being pulled close to each other. In this book, however, we will not be very strict in this respect because the very definition of relative phase between two given oscillators, especially when they represent oscillators of a different nature, is rather arbitrary. The importance of the function of synchronization in the self-organization in nature may be realized from the fact that what looks like a single periodic process on a macroscopic level often turns out to be a collective oscillation resulting from the mutual synchronization among the tremendous number of the constituent oscillators. The human heartbeat may serve as an example of such a phenomenon. Because the component oscillators in nature would never possess identical natural frequencies and, moreover, would never be free from environmental random fluctuations, mutual synchronization appears to be the unique possible mechanism for producing and maintaining macroscopic rhythmicity. The problem of the onset of collective oscillation in oscillator aggregates will be treated in Chap. 5.

Based equally on mutual synchronization, chemical wave propagation in oscillatory reaction-diffusion systems generates an even more complicated class of phenomena than the mere collective oscillations. Here again, the entire field may be entrained into an identical frequency, whereas the local phases of oscillation may have different values. Such a view, although a little too idealized, enables us to understand the origin of expanding target patterns as observed in the Belousov-Zhabotinsky reaction. As implied from a number of problems in condensed-matter physics and field theory, some field quantity for which the phase is definable is expected to allow for topological defects arising from, e.g., phase jumps and phaseless points. Just as in superfluid helium and plane rotator systems, there exist in oscillatory reaction-diffusion systems, too, vortexlike modes which, in the latter case, develop into rotating spiral waves such as those known in the Belousov-Zhabotinsky reaction. A simple theory of chemical waves from the viewpoint of spatio-temporal synchronization and phase singularity will be developed in Chap. 6. Granted that mutual synchronization represents a key mechanism in the self-organization in oscillatory media, it would be interesting to ask what is brought about by its breakdown. This will partly be answered in Chap. 7 where we try to relate it to the onset of turbulencelike behavior.

It is unfortunate that only little progress has been achieved in the past towards the understanding of synchronization, pattern formation, and turbulence in nonlinear self-oscillatory media and related many-body systems, in spite of their great potential importance in the future. Although the present theory, too, is far from complete, a particular physical viewpoint at least will be seen to underlie the present book. In Part 1, such a viewpoint will be formulated into some asymptotic methods, while Part 2 may simply be looked upon as their demonstration through a number of specific problems. The underlying physics is closely related to the slaving principle, whose conceptual importance in nonlinear dissipative dynamics in general was emphasized by Haken and first demonstrated by him in laser theory (Haken and Sauermann, 1963; Haken, 1983a, b). Basically, the slaving principle claims the possibility of eliminating a large number of rapidly decaying degrees of freedom. This principle manifests itself most clearly near the bifurcation points where the system experiences a qualitative change in dynamical behavior. The possibility of a great reduction of the number of effective degrees of freedom and the resulting universality of the evolution law form the physical basis of why the bifurcation theory can serve as a most powerful tool in treating various self-organization phenomena.

It should be noted, however, that the slaving principle is such a general concept that the standard bifurcation theory can embody only a part of this concept on a more or less firm mathematical basis. Thus, the theory developed in this book, although being based on the slaving principle, is not so much based on the standard bifurcation theory. In fact, the kinds of self-organization phenomena and turbulence we want to treat here are rather complicated and require so many effective degrees of freedom that standard bifurcation theory does not seem to be of much help. As a possible alternative, one may think of the bifurcation theory of higher codimensions, which has shown an interesting development in recent years (e.g, Guckenheimer, 1981). Unfortunately, however, the effective degrees of freedom involved are still too few for our purposes. We are rather interested in, so to speak, the bifurcations with *infinitely* high degeneracy, which can in fact cover some physical problems of our concern. Although no rigorous bifurcation theory seems available for such cases, this peculiar kind of bifurcation is of much practical importance. This is because it arises quite commonly in systems with great spatial extension, especially when the instability first occurs for disturbances of sufficiently long spatial scales. As inferred from the fact that the eigenvalue spectrum of the fluctuations around the subcritical state is then almost continuous, the system dynamics can never be confined to a few-dimensional manifold even in the vicinity of the bifurcation point. Or it may be better to say that if the range of applicability of the usual bifurcation theory is measured by some bifurcation parameter, it will be narrowed down to zero as the system extension goes to infinity. Even in such highly degenerate bifurcations, there exist a tremendous number of degrees of freedom which are rapidly decaying and hence follow adiabatically the continuum of long-scale modes. Thus, the idea of the slaving principle itself is expected to work. Although lacking a rigorous mathematical basis, some practical methods of dynamical reduction appropriate for highly degenerate bifurcations were developed in some fluid-dynamical problems such as the plane Poiseuille flow (Stewartson and Stuart, 1971) and the Rayleigh-Bénard convection (Newell and Whitehead, 1969). In Chap. 2 we apply this kind of approach to simpler systems, i.e., oscillatory reaction-diffusion systems, with a special emphasis on its formalistic contrast to the ordinary Hopf bifurcation theory.

The utility of the slaving principle is by no means restricted to near-bifurcation situations. In fact, in connection with the present concern, the slaving principle is also applicable to systems of *weakly* coupled limit cycle oscillators in general. A theoretical framework particularly suited for weakly coupled oscillators and systems analogous to them will be presented in Chaps. 3 and 4. Here the simplification of the dynamics comes essentially from the fact that the amplitude disturbances decay much faster than the phase disturbances. In conclusion, the slaving principle enables us to contract the original dynamics to a much simpler one which still retains a sufficiently large number of effective degrees of freedom to admit a variety of self-organization and turbulent phenomena.

Finally, one should keep in mind the severe limitation of the present methods in that they apply only to those phenomena associated with sufficiently long space-time scales. It is under this restriction that the present theory can enjoy its coherent character. A number of important phenomena in self-oscillating fields, especially those for which the coexistence of short-scale and long-scale spatial variations are important, are omitted. An important problem of this kind would be the propagation of trigger waves in reaction-diffusion systems, i.e., waves which are typical in non-oscillating excitable media but may arise also in systems of highly distorted oscillations or relaxation oscillations. Some simple classes of phenomena related to trigger waves may be dealt with by a method quite different from the present ones, for which the reader may refer to Ortoleva and Ross (1975) and Fife (1976a, b, 1979b).

2. Reductive Perturbation Method

Small-amplitude oscillations near the Hopf bifurcation point are generally governed by a simple evolution equation. If such oscillators form a field through diffusion-coupling, the governing equation is a simple partial differential equation called the Ginzburg-Landau equation.

2.1 Oscillators Versus Fields of Oscillators

Many theories on the nonlinear dynamics of dissipative systems are based on the first-order ordinary differential equations

$$\frac{dX_i}{dt} = F_i(X_1, X_2, \ldots, X_n; \mu), \quad i = 1, 2, \ldots, n,$$

which include some parameters represented by μ; a more convenient vector form

$$\frac{d\boldsymbol{X}}{dt} = \boldsymbol{F}(\boldsymbol{X}; \mu) \tag{2.1.1}$$

is sometimes preferred. As a specific example, we mention the dynamics of chemical reaction systems which are maintained uniformly in space. In this case, $\boldsymbol{X}$ usually represents a set of concentrations of the chemical species involved, and μ may be taken to be the flow rate at which certain chemicals are constantly fed into the system so that their consumption due to reactions may be compensated.

For some range of μ, the system may stay stable in a time-independent state. In particular, this is usually the case for macroscopic physical systems which lie sufficiently close to thermal equilibrium. In many systems, such a steady state loses stability at some critical value μ_c of μ, and beyond it (say $\mu > \mu_c$), gives way to periodic motion. In the parameter-amplitude plane, this appears as a branching of time-periodic solutions from a stationary solution branch, and this phenomenon is generally called the *Hopf bifurcation*. For various mathematical aspects of the Hopf bifurcation, one may refer to the book by Marsden and McCracken (1976). In chemical reactions, the corresponding phenomenon is called the onset of *chemical oscillations*. Besides chemical reactions, one may point out many examples from electrical and mechanical engineering, optics, biology, biochemistry, and possibly some other fields, for which ordinary-differential-equation models form a natural basis for mathematical analysis, so that the appearance of oscillations may be understood in the way stated above.

As μ increases further, the system may show more and more complicated dynamics through a number of bifurcations. It may show complicated periodic oscillations, quasi-periodic oscillations or a variety of non-periodic behaviors. For instance, we know of the recent discoveries of fantastic bifurcation structures in the spatially homogeneous Belousov-Zhabotinsky reaction, see Hudson et al., 1979.

Coming back to limit cycle oscillations shown by systems of ordinary differential equations, this simple mode of motion still seems to deserve some more attention, especially in relation to its role as a basic functional unit from which various dynamical complexities arise. This seems to occur in at least two ways. As mentioned above, one may start with a simple oscillator, increase μ, and obtain complicated behaviors; this forms, in fact, a modern topic. However, another implication of this dynamical unit should not be left unnoticed. We should know that a limit cycle oscillator is also an important component system in various self-organization phenomena and also in other forms of spatio-temporal complexity such as turbulence. In this book, particular emphasis will be placed on this second aspect of oscillator systems. This naturally leads to the notion of the "many-body theory of limit cycle oscillators"; we let many oscillators contact each other to form a "field", and ask what modes of self-organization are possible or under what conditions spatio-temporal chaos arises, etc. A representative class of such many-oscillator systems in theory and practical application is that of the fields of diffusion-coupled oscillators (possibly with suitable modifications), so that this type of system will primarily be considered in this book.

In any case, we should begin with some investigation of the component systems, i.e., limit cycle oscillators. Although the specific feature of limit cycle oscillations (e.g., orbital forms, oscillation patterns, etc.) may vary greatly from system to system, there exists one remarkable universal fact, namely, that all systems come to behave in a similar manner sufficiently close to the onset of oscillations. Mathematicians may say that this is a consequence of the *center manifold theorem*. More physically, we are left with only a couple of relevant dynamical variables close to criticality, whose time scales are distinguishably slower than those of the remaining variables, so that the latter can be eliminated adiabatically. As a result, (2.1.1) is contracted to a very simple universal equation which is sometimes called the Stuart-Landau equation. In fact, Landau was the first to conjecture the equation form (Landau, 1944), and Stuart was the first to derive it through an asymptotic method (Stuart, 1960). In quite a different context, specifically in laser theory, Haken and Sauermann (1963) derived a similar but more general equation. We shall outline in Sect. 2.2 how the Stuart-Landau equation is derived. The fact that dynamical systems can be reduced to some simple universal systems is by no means restricted to this particular bifurcation type. However, we do not intend in this book to present theories from such a general viewpoint. The method employed in Sect. 2.2 is a well-known multi-scale method, although there may be some possible variants leading to identical results. A practical use of the theory in Sect. 2.2 lies in the fact that it enables us to calculate explicitly a certain constant (called the Landau constant) appearing in the Stuart-Landau equation, whose sign determines the stability of the bifurcating periodic solution. Otherwise, the Stuart-Landau equation itself is not likely to arouse much theoretical interest, although it may have some value in serving as an ideal nonlinear oscillator model.

So far, the discussion has been concerned with systems of *ordinary* differential equations. In many physical problems, *partial* differential equations describing processes in the space-time domain prove to be a more useful mathematical tool. For instance, one may mention the Navier-Stokes fluids, chemical reactions including diffusion, some ecological systems with migration, etc. Suppose that oscillatory motions occur in any of these continuous media as some control parameter is varied, and consider how to describe them. It is true that if the system is confined within a finite volume, the governing partial differential equations can, in principle, be transformed into a discrete set of ordinary differential equations, which describe the evolution of the amplitudes of the basis functions satisfying prescribed boundary conditions. Although the system then involves an infinite number of degrees of freedom, a mode-truncation approximation is usually allowed. Thus, as far as the onset of oscillations is concerned, there seems to be nothing theoretically new, compared to the bifurcation theory for systems of ordinary differential equations. Specifically, the application of a multi-scale method will lead to a Stuart-Landau equation again. (For a mathematical theory of the Hopf bifurcation for systems of partial differential equations in bounded domains, see Joseph and Sattinger, 1972; bifurcation analyses of reaction-diffusion systems have been developed by Auchmuty and Nicolis, 1975, 1976, and Herschkowitz-Kaufman, 1975.)

There may be some situations, however, where keeping to formal bifurcation theories easily makes us overlook a fact of considerable physical importance. The situation of particular interest in this connection seems to be when the system size is very large. Then, formal bifurcation techniques applied near μ_c cannot claim full validity except in an extremely limited parameter range about μ_c. This is basically because the eigenvalue spectrum obtained from the linearization about the reference steady state is almost continuous for large system size, so that, in addition to the couple of modes which are becoming unstable, a large number of degrees of freedom come into play as soon as μ deviates from μ_c (a more detailed description will be given in Sect. 2.3). Thus it is desirable that the Stuart-Landau equation be generalized so as to cover such circumstances. People in the field of fluid mechanics have developed theories in this direction, which proved to be very useful in understanding instabilities (not restricted to the Hopf type) arising in systems with large dimensions at least in one or two directions. Typical examples are the Newell-Whitehead theory (1969) on a fluid layer heated from below with infinite aspect ratio, and the Stewartson-Stuart theory (1971) on plane Poiseuille flow. In these theories, one works with partial differential equations throughout, not transforming them into ordinary differential equations. A method was contrived to reduce the equations to a generalized form of the Stuart-Landau equation, thereby admitting slow spatial and temporal modulation of the envelope of the bifurcating flow patterns. We call that equation the Ginzburg-Landau equation (named after a similar equation appearing in superconductivity) or the Stewartson-Stuart equation. In this book we adopt the former name.

Independently of the hydrodynamical context, the Ginzburg-Landau equation was derived by Graham and Haken (1968, 1970) in multimode lasers as a further development of the Haken-Sauermann theory (1963); it should be noted that fluctuations are included in most of their series of works. For various non-equilibrium phase transitions described by the Ginzburg-Landau-type equation, see the review article by Haken (1975 b) and his more recent monograph (1983).

The derivation of the Ginzburg-Landau equation usually involves the method of multiple scales (in space and time), and again there are some variants in technical details. For convenience, we sometimes call all the related techniques involving the use of stretched space-time coordinates *the reductive perturbation method*, a term originally coined for a systematic method of deriving various nonlinear wave equations mainly in dissipationless media (Taniuti and Wei, 1968; Taniuti, 1974). It is now widely known that the Ginzburg-Landau equation is not only related to a few fluid mechanical or optical problems but that it can be deduced from a rather general class of partial differential equations (Newell, 1974; Haken, 1975a; Gibbon and McGuiness, 1981; Lin and Kahn, 1982). Chemical reactions with diffusion form a simple and particularly interesting class of systems in this connection (Kuramoto and Tsuzuki, 1974; Wunderlin and Haken, 1975), and we shall derive in Sect. 2.4 the Ginzburg-Landau equation for general reaction-diffusion systems. Just as the Stuart-Landau equation describes the simplest nonlinear oscillator, so the Ginzburg-Landau equation describes the simplest *field* of nonlinear oscillators. In later chapters, this equation will be frequently invoked in discussing chemical waves and chemical turbulence.

2.2 The Stuart-Landau Equation

In this section, we outline how a small-amplitude equation valid near the Hopf bifurcation point is derived from the general system of ordinary differential equations (2.1.1).

Let X and F be n-dimensional real vectors and μ a real scalar parameter. Let $X_0(\mu)$ denote a steady solution of (2.1.1) or

$$F(X_0(\mu);\mu) = 0 .$$

We express (2.1.1) in terms of the deviation $u \equiv X - X_0$ in a Taylor series:

$$\frac{du}{dt} = Lu + Muu + Nuuu + \dots , \qquad (2.2.1)$$

where L denotes the Jacobian matrix whose ijth element is given by $L_{ij} = \partial F_i(X_0)/\partial X_{0j}$; the abbreviations Muu and $Nuuu$, etc., indicate vectors whose ith components are given by

$$(\boldsymbol{Muu})_i = \sum_{j,k} \frac{1}{2!} \frac{\partial^2 F_i(\boldsymbol{X}_0)}{\partial X_{0j} \partial X_{0k}} u_j u_k ,$$

$$(\boldsymbol{Nuuu})_i = \sum_{j,k,l} \frac{1}{3!} \frac{\partial^3 F_i(\boldsymbol{X}_0)}{\partial X_{0j} \partial X_{0k} \partial X_{0l}} u_j u_k u_l ,$$

and the higher-order terms in $\boldsymbol{u}$ may be expressed similarly. We shall later use quantities like $\boldsymbol{Muv}$ and $\boldsymbol{Nuvw}$ for different vectors $\boldsymbol{u}$, $\boldsymbol{v}$ and $\boldsymbol{w}$, and their definitions may be understood as an obvious extension of the above. Note, in particular, that $\boldsymbol{Muv}$ and $\boldsymbol{Nuvw}$ are symmetric functions of $\boldsymbol{u}$, $\boldsymbol{v}$ and $\boldsymbol{w}$. Note also that the expansion coefficients, which are symbolically expressed by M, N, etc., generally depend on μ at least through $X_0(\mu)$.

Suppose that μ is varied in some range about $\mu = 0$. We assume that up to $\mu = 0$ the solution $\boldsymbol{X}_0$ remains stable to sufficiently small perturbations, while it loses stability for $\mu > 0$. Consider the linear eigenvalue problem associated with (2.2.1), i.e.,

$$L\boldsymbol{u} = \lambda \boldsymbol{u} . \tag{2.2.2}$$

The stability of $\boldsymbol{X}_0$ is related to the distribution of the eigenvalues λ in the complex plane. By assumption, this distribution changes with μ in the following way: all λ stay in the left half-plane if $\mu < 0$, and at least one eigenvalue crosses the imaginary axis at $\mu = 0$. Since the eigenvalues are given by the zeros of an nth-order polynomial with real coefficients, we have the following two general possibilities: (a) one eigenvalue on the real axis crosses the origin (Fig. 2.1 a), (b) a pair of complex-conjugate eigenvalues cross the imaginary axis simultaneously (Fig. 2.1 b). In each case, the eigenvalues are assumed to have nonzero transversal "velocity" when crossing the imaginary axis, or

$$\left. \frac{d\,\mathrm{Re}\{\lambda(\mu)\}}{d\mu} \right|_{\mu=0} > 0 .$$

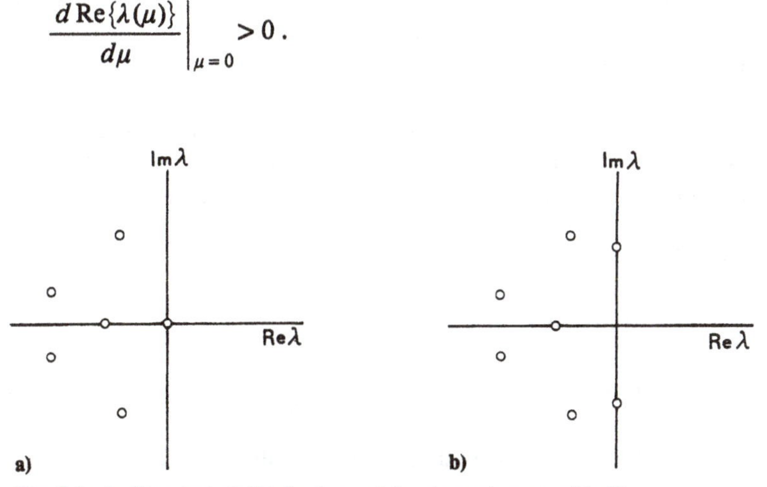

Fig. 2.1 a, b. Two typical distributions of the eigenvalues at criticality

Furthermore, the rest of the eigenvalues are assumed to remain at a nonzero distance from the imaginary axis. In the following, we shall restrict our attention to case (b), since this corresponds to the Hopf bifurcation.

Near criticality, the matrix L may be developed in powers of μ:

$$L = L_0 + \mu L_1 + \mu^2 L_2 + \dots \, . \tag{2.2.3}$$

To save notation, let $\lambda(\mu)$ denote a special eigenvalue which is becoming critical rather than denoting a general one, and $\bar{\lambda}(\mu)$ its complex conjugate (we use a bar to signify a complex conjugate throughout). We assume a power-series expansion for λ also:

$$\lambda = \lambda_0 + \mu\lambda_1 + \mu^2\lambda_2 + \dots \, , \tag{2.2.4}$$

where λ_ν are generally complex, or $\lambda_\nu = \sigma_\nu + \mathrm{i}\,\omega_\nu$. By assumption,

$$\sigma_0 = 0 \, , \qquad \sigma_1 > 0 \, .$$

Let U denote the right eigenvector of L_0 corresponding to the eigenvalue $\lambda_0(=\mathrm{i}\,\omega_0)$:

$$L_0 U = \lambda_0 U \, , \qquad L_0 \bar{U} = \bar{\lambda}_0 \bar{U} \, .$$

Similarly, the left eigenvector is denoted by U^*:

$$U^* L_0 = \lambda_0 U^* \, , \qquad \bar{U}^* L_0 = \bar{\lambda}_0 \bar{U}^* \, ,$$

where $U^*\bar{U} = \bar{U}^* U = 0$, and these vectors are normalized as $U^* U = \bar{U}^* \bar{U} = 1$. Note that λ_0 and λ_1 are expressed as

$$\lambda_0 = \mathrm{i}\,\omega_0 = U^* L_0 U \, , \tag{2.2.5 a}$$

$$\lambda_1 = \sigma_1 + \mathrm{i}\,\omega_1 = U^* L_1 U \, . \tag{2.2.5 b}$$

It is convenient to define a small positive parameter ε by $\varepsilon^2\chi = \mu$, where $\chi = \operatorname{sgn}\mu$; ε is considered to be a measure of the amplitude to lowest order, so that one may assume the expansion

$$\boldsymbol{u} = \varepsilon \boldsymbol{u}_1 + \varepsilon^2 \boldsymbol{u}_2 + \dots \, . \tag{2.2.6}$$

The expression in (2.2.3) now becomes

$$L = L_0 + \varepsilon^2 \chi L_1 + \varepsilon^4 L_2 + \dots \, . \tag{2.2.7}$$

Similarly, for some higher-order expansion coefficients in (2.2.1), we write symbolically

$$M = M_0 + \varepsilon^2 \chi M_1 + \dots\,,$$
$$N = N_0 + \varepsilon^2 \chi N_1 + \dots\,. \tag{2.2.8}$$

From the fact that λ has a small real part of order ε^2, it would be appropriate to introduce a scaled time τ via

$$\tau = \varepsilon^2 t\,,$$

and regard $\boldsymbol{u}$ as depending both on t and τ, and having no explicit dependence on ε; t and τ will be treated as mutually independent. Correspondingly, the time differentiation in (2.2.1) should be transformed as

$$\frac{d}{dt} \to \frac{\partial}{\partial t} + \varepsilon^2 \frac{\partial}{\partial \tau}\,. \tag{2.2.9}$$

The substitution of (2.2.6 – 9) into (2.2.1) gives

$$\left(\frac{\partial}{\partial t} + \varepsilon^2 \frac{\partial}{\partial \tau} - L_0 - \varepsilon^2 \chi L_1 - \dots\right)(\varepsilon \boldsymbol{u}_1 + \varepsilon^2 \boldsymbol{u}_2 + \dots)$$
$$= \varepsilon^2 M_0 \boldsymbol{u}_1 \boldsymbol{u}_1 + \varepsilon^3 (2 M_0 \boldsymbol{u}_1 \boldsymbol{u}_2 + N_0 \boldsymbol{u}_1 \boldsymbol{u}_1 \boldsymbol{u}_1) + O(\varepsilon^4)\,. \tag{2.2.10}$$

Equating coefficients of different powers of ε in (2.2.10), we have a set of equations in the form

$$\left(\frac{\partial}{\partial t} - L_0\right) \boldsymbol{u}_\nu = \boldsymbol{B}_\nu\,, \qquad \nu = 1, 2, \dots\,. \tag{2.2.11}$$

The first few $\boldsymbol{B}_\nu$ are

$$\boldsymbol{B}_1 = 0\,, \tag{2.2.12a}$$

$$\boldsymbol{B}_2 = M_0 \boldsymbol{u}_1 \boldsymbol{u}_1\,, \tag{2.2.12b}$$

$$\boldsymbol{B}_3 = -\left(\frac{\partial}{\partial \tau} - \chi L_1\right) \boldsymbol{u}_1 + 2 M_0 \boldsymbol{u}_1 \boldsymbol{u}_2 + N_0 \boldsymbol{u}_1 \boldsymbol{u}_1 \boldsymbol{u}_1\,. \tag{2.2.12c}$$

In general, the $\boldsymbol{B}_\nu$ are functions of the lower-order quantities $\boldsymbol{u}_{\nu'}$ $(\nu' < \nu)$.

For the system of linear inhomogeneous equations (2.2.11), we have an important property:

$$\int_0^{2\pi/\omega_0} \boldsymbol{U}^* \cdot \boldsymbol{B}_\nu \mathrm{e}^{-\mathrm{i}\omega_0 t} dt = 0\,, \tag{2.2.13}$$

which follows from

$$\int_0^{2\pi/\omega_0} U^* \cdot B_\nu e^{-i\omega_0 t} dt = \int_0^{2\pi/\omega_0} \left[U^* \cdot \left(\frac{\partial}{\partial t} - L_0 \right) u_\nu \right] e^{-i\omega_0 t} dt$$
$$= \int_0^{2\pi/\omega_0} (i\omega_0 U^* \cdot u_\nu - i\omega_0 U^* \cdot u_\nu) e^{-i\omega_0 t} dt = 0 .$$

The equality (2.2.13) is called the solvability condition. By inspecting the general structure of (2.2.11), it is expected that the u_ν can be found iteratively as 2π-periodic functions of $\omega_0 t$. This means that $B_\nu(t, \tau)$ is also 2π-periodic in $\omega_0 t$, so that it would be appropriate to express it in the form

$$B_\nu(t, \tau) = \sum_{l=-\infty}^{\infty} B_\nu^{(l)}(\tau) e^{il\omega_0 t} .$$

The solvability condition now reduces to

$$U^* \cdot B_\nu^{(1)}(\tau) = 0 , \tag{2.2.14}$$

which is the crucial condition used below.

For $\nu = 1$, (2.2.11) may be solved in the form

$$u_1(t, \tau) = W(\tau) U e^{i\omega_0 t} + \text{c.c.} , \tag{2.2.15}$$

where c.c. stands for the complex conjugate, and $W(\tau)$ is some complex amplitude yet to be specified. This is called the neutral solution. We shall soon find later that the evolution equation for W, which is nothing but the Stuart-Landau equation, is given by (2.2.14) for $\nu = 3$, or by

$$U^* \cdot B_3^{(1)} = 0 . \tag{2.2.16}$$

Note that (2.2.14) is trivially satisfied for $\nu = 2$ because $B_2^{(1)}$ vanishes identically as is clear from (2.2.12b) and (2.2.15). In order to derive the equation obeyed by W, it is thus necessary to express u_2 appearing in $B_3^{(1)}$ in terms of u_1 (or W), and this can be done by solving (2.2.11) for u_2 as a function of u_1. But B_2 contains only the zeroth and second harmonics, and the same is true for u_2. Thus, we try to find u_2 in the form

$$u_2 = V_+ W^2 e^{2i\omega_0 t} + V_- \bar{W}^2 e^{-2i\omega_0 t} + V_0 |W|^2 + v_0 u_1 . \tag{2.2.17a}$$

By substituting this into (2.2.11) for $\nu = 2$, the quantities $V_{\pm,0}$ are obtained in the form

$$\begin{aligned} V_+ &= \bar{V}_- = -(L_0 - 2i\omega_0)^{-1} M_0 U U , \\ V_0 &= -2 L_0^{-1} M_0 U \bar{U} . \end{aligned} \tag{2.2.17b}$$

The constant v_0 cannot be determined at this stage,but we do not need it for the present purpose. Substituting (2.2.15, 17a) into (2.2.12c), we have

$$B_3^{(1)} = -\left(\frac{\partial}{\partial \tau} - \chi L_1\right) WU + (2M_0 U V_0 + 2M_0 \bar{U} V_+ + 3N_0 U U \bar{U}) |W|^2 W. \tag{2.2.18}$$

Then the solvability condition (2.2.16) itself turns out to take the form of the Stuart-Landau equation

$$\frac{\partial W}{\partial \tau} = \chi \lambda_1 W - g|W|^2 W, \tag{2.2.19}$$

where g is a complex number given by

$$g \equiv g' + \mathrm{i}g'' = -2U^* M_0 U V_0 - 2U^* M_0 \bar{U} V_+ - 3U^* N_0 U U \bar{U}. \tag{2.2.20}$$

Defining the amplitude R and the phase Θ via $W = R\exp(\mathrm{i}\,\Theta)$, one may alternatively write (2.2.19) as

$$\begin{aligned} \frac{dR}{d\tau} &= \chi\sigma_1 R - g' R^3, \\ \frac{d\Theta}{d\tau} &= \chi\omega_1 - g'' R^2. \end{aligned} \tag{2.2.21}$$

The non-trivial solution

$$R = R_s, \qquad \Theta = \tilde{\omega} t + \text{const},$$

$$R_s = \sqrt{\sigma_1/|g'|}, \qquad \tilde{\omega} = \chi(\omega_1 - g'' R_s^2),$$

appears only in the supercritical region ($\chi > 0$) for positive g' and in the subcritical region for negative g'. In the former case, the bifurcation is called supercritical, and in the latter case, subcritical. Linearization about R_s shows that the supercritical bifurcating solution is stable, while the subcritical one is unstable. The bifurcating solution shows a perfectly smooth circular motion in the complex W plane. The corresponding expression for the original vector variable X is approximately given by

$$X \simeq X_0 + \varepsilon u_1 = X_0 + \varepsilon\{U R_s \exp[\mathrm{i}(\omega_0 + \varepsilon^2 \tilde{\omega})t] + \text{c.c.}\},$$

which describes a small-amplitude elliptic orbital motion in the critical eigenplane.

2.3 Onset of Oscillations in Distributed Systems

The foregoing argument was about systems of ordinary differential equations. For chemical reactions this corresponds to the dynamics of local systems which are supposed to be uncoupled from the surroundings or, otherwise, the dynamics of the entire system, which is kept uniform, e.g., by continuous stirring. We wish to consider more general circumstances where the system is left unstirred so that non-uniform concentration fluctuations may arise. An appropriate mathematical model is then obtained simply by adding diffusion terms to (2.1.1) as

$$\frac{\partial \boldsymbol{X}}{\partial t} = \boldsymbol{F}(\boldsymbol{X}) + D\nabla^2 \boldsymbol{X} . \tag{2.3.1}$$

This is called a reaction-diffusion equation; D is a matrix of diffusion constants, and is often assumed to be diagonal. The specific problem we shall consider in the rest of this chapter concerns the possible roles of the many spatial degrees of freedom taken on when a uniform steady state experiences an instability of oscillatory type. In this section, we give only a qualitative argument; the formal derivation of the Ginzburg-Landau equation is rather easy, and will be done in the next section.

To make the point clear, we shall work in this section with the system (2.3.1) in an interval $-\xi/2 \leq x \leq \xi/2$, subject to the no-flux boundary conditions

$$\left.\frac{\partial \boldsymbol{X}}{\partial x}\right|_{\pm\xi/2} = 0 . \tag{2.3.2}$$

The stability of the uniform steady solution $\boldsymbol{X}_0$ of (2.3.1) may then be analyzed from the variational equations about $\boldsymbol{X}_0$,

$$\frac{\partial \boldsymbol{u}}{\partial t} = \left(L + D\frac{\partial^2}{\partial x^2}\right)\boldsymbol{u} . \tag{2.3.3}$$

For any integer l, this equation admits solutions of the form

$$\boldsymbol{u}(x,t) = \boldsymbol{V}_l \mathrm{e}^{\lambda t} \cos\frac{l\pi}{\xi}x .$$

By substitution, we have the eigenvalue problem

$$L_l \boldsymbol{V}_l = \lambda \boldsymbol{V}_l ,$$
$$L_l = L - D\left(\frac{l\pi}{\xi}\right)^2 , \tag{2.3.4}$$

which reduces for vanishing l to the eigenvalue problem (2.2.2) already quoted for uniform systems. Each of these previous eigenvalues now forms a branch which is almost continuous if the system length ξ is very large (Fig. 2.2); specifically, the distance between two neighboring eigenvalues in the same branch is of order $D_0\xi^{-2}$ if $l \sim O(1)$, where D_0 is the typical magnitude of the diffusion constants, hereafter set equal to 1. We assume that up to $\mu = 0$ all eigenvalues lie

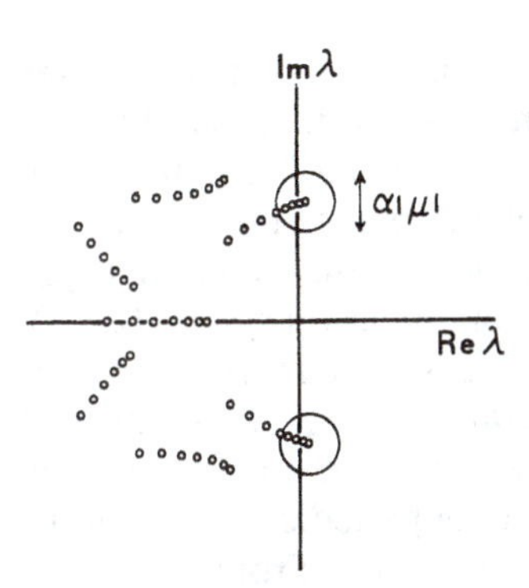

Fig. 2.2. Schematic distribution of the eigenvalues for large systems. Only the eigenmodes corresponding to the eigenvalues inside the circles are relevant to the dynamics near criticality

in the left half-plane and, as before, a pair of complex-conjugate eigenvalues corresponding to $l=0$ cross the imaginary axis at $\mu=0$. Note that in assuming this, we exclude the possibility of $\boldsymbol{X}_0$ becoming unstable to *nonuniform* fluctuations, although this type of instability, known as the diffusion instability (of the Rashevsky-Turing type), could be of considerable theoretical interest (Rashevsky, 1940; Turing, 1952; for an extensive study of *dissipative structure* arising from this type of instability, see Nicolis and Prigogine, 1977, and references therein). With this condition, we wish to know if it is still possible that non-uniform modes have some important effects on the dynamics close to criticality. The answer seems to depend on the relative size of μ and ξ^{-2}. To see this, we first decompose $\boldsymbol{u}(x,t)$ into eigenmodes:

$$\boldsymbol{u}(x,t)=\sum_{\alpha,l} c_{\alpha l}(t)\, \boldsymbol{V}_{\alpha l}\cos\frac{l\pi}{\xi}x\,,$$

where $\boldsymbol{V}_{\alpha l}(\alpha=1,2,\ldots,n)$ denote the eigenvectors of L_l. Then our reaction-diffusion system (2.3.1) reduces to a system of ordinary differential equations as

$$\frac{dc_{\alpha l}}{dt}=\lambda_{\alpha l}c_{\alpha l}+\text{nonlinear terms}\,. \tag{2.3.5}$$

We are now working with an infinite number of variables $c_{\alpha l}$ in contrast to the previous system of ordinary differential equations. However, the mode amplitudes $c_{\alpha l}$ for sufficiently large l may safely be neglected since they decay strongly, so that the asymptotic method comes to work again. However, the application of this method leads to the result that $c_{\alpha l}=0$ for all non-vanishing l. This means that the system remains uniform, and we trivially come back to the theory of Sect. 2.2. The reason why the non-uniform modes were entirely eliminated is that we took the limit $\mu\to 0$ under fixed ξ, however large the latter may be. There may be some non-uniform fluctuations initially, but they will decay very rapidly compared to the critical mode amplitudes. In fact, the characteristic time scale of the slowest non-uniform modes is roughly estimated to be $\tau_1\sim(\mu+\xi^{-2})^{-1}$, while for the critical modes it is $\tau_0\sim\mu^{-1}$. Although both these could be very long, the ratio τ_1/τ_0 goes to zero as $\mu\to 0$ for finite ξ. Moreover, our perturbation theory has been so formulated that such initial transients are automatically eliminated (possibly by restricting the class of initial conditions to the center manifold).

It is clear that the practical value of any asymptotic theory lies in the applicability of the resulting formula to finite values of the parameters, which are supposed to be infinitesimal in the theory. Inquiring into the range of applicability of the Stuart-Landau equation for the reaction-diffusion system (2.3.1), we find it must be limited to $\tau_1/\tau_0 \ll 1$ or $|\mu| \ll \xi^{-2}$. This is indeed very narrow for a very large system size ξ. Once $|\mu|$ becomes comparable with ξ^{-2}, a number of wavelike modes will come to have time scales comparable to those of the critical modes, and then even the stability of the bifurcating time-periodic solution of the supercritical Stuart-Landau equation becomes questionable. From this argument, the reason is now clear why we have to develop a generalized perturbation method for a large system size so that some non-uniform modes may be accommodated even in the neighborhood of the bifurcation point.

The above discussion also suggests that there are in the (μ, ξ) parameter plane three distinct characteristic regimes which are $|\mu| \ll \xi^{-2}$, $|\mu| \sim \xi^{-2}$, and $|\mu| \gg \xi^{-2}$; in these cases, the number of non-uniform modes to be taken into account is null, a few, and very many, respectively. In order to make the argument a little more precise, it is appropriate to put

$$\xi = \xi_0 |\mu|^{-\delta},$$

and consider the limit $\mu \to 0$. One may classify various asymptotic regimes according to δ:

$$\text{(A)}\ \delta > \tfrac{1}{2}, \quad \text{(B)}\ \delta = \tfrac{1}{2}, \quad \text{(C)}\ \delta < \tfrac{1}{2}.$$

Their physical distinction may be seen as follows: In Fig. 2.2 we see a pair of small circles in the λ plane centered at the eigenvalues which are becoming critical. Their radius is taken to be $\alpha|\mu|$, where α is some constant independent of μ or ξ. Then, it is clear that the total number n of eigenvalues lying in the encircled regions becomes

$$\text{(A)}\ n = \infty, \quad \text{(B)}\ 2 \le n < \infty\ (\text{depends on } \alpha \text{ and } \xi_0), \quad \text{(C)}\ n = 2,$$

as $\mu \to 0$. Since all modes in the circles have comparable growth or decay rates, they have to be treated as equally important; consequently, the contracted form of the evolution equation to be derived near criticality must allow for these degrees of freedom. It seems, however, unreasonable to put a clear borderline at such circles separating slower and faster modes; if we persist in doing this, α should safely be taken to be sufficiently large. To be more precise, we have first to let μ and ξ^{-1} tend to zero, and then let α go to infinity. Thus the list for n above should be modified to

$$\text{(A) (B)}\ n = \infty, \quad \text{(C)}\ n = 2.$$

This implies that the contracted form of the dynamics will be obtained as partial differential equations for (A) and (B), and ordinary differential equations (specifically, the Stuart-Landau equation) for (C). In both cases, the encircled part of the eigenmodes is an infinitesimal portion of the total number of eigenmodes, so that a great reduction of the degrees of freedom is going to be achieved. One may alternatively say that a notion something like the center manifold is still working even for (A) and (B), although its dimension could no longer be finite.

2.4 The Ginzburg-Landau Equation

We now concentrate attention on the reaction-diffusion system (2.3.1). The system dimension and size, geometrical form, and boundary conditions imposed are not specified at first. In particular, the theory to be developed below applies to situations (A) – (C) equally well. The formulation goes quite in parallel with that in Sect. 2.2. We express the reaction-diffusion equations in terms of $\boldsymbol{u}(\boldsymbol{r},t)$, the space-time dependent deviation from the uniform steady solution $\boldsymbol{X}_0$:

$$\frac{\partial \boldsymbol{u}}{\partial t} = (L + D\nabla^2)\boldsymbol{u} + M\boldsymbol{u}\boldsymbol{u} + N\boldsymbol{u}\boldsymbol{u}\boldsymbol{u} + \dots \, . \tag{2.4.1}$$

Previously, we regarded $\boldsymbol{u}$ as dependent on two time scales, t and τ. We now also allow for its slow space dependence characterized by the slowness parameter $\varepsilon(=|\mu|^{1/2})$, and we absorb this extra ε dependence into a scaled coordinate $\boldsymbol{s}$, defined by

$$\boldsymbol{s} = \varepsilon \boldsymbol{r} \, . \tag{2.4.2}$$

In this way, we regard $\boldsymbol{u}$ as a function of t, τ, and $\boldsymbol{s}$. The reason why we should assume this slow space dependence is clear from the discussion in the previous section. To put it briefly, we are treating the circumstance where the long-wavelength modes inside the circles in Fig. 2.2 are generally present with nonvanishing amplitudes, while the shorter-wavelength modes outside the circles are absent.

In addition to the transformation in (2.2.9), we have to make in (2.4.1) the transformation

$$\nabla \to \varepsilon \nabla_s \, . \tag{2.4.3}$$

We let L, M, N, etc., be developed in powers of ε in the form of (2.2.7, 8). They are substituted into (2.4.1), and after making transformations (2.2.9) and (2.4.3), we have

$$\left(\frac{\partial}{\partial t} + \varepsilon^2 \frac{\partial}{\partial \tau} - \varepsilon^2 D \nabla_s^2 - L_0 - \varepsilon^2 \chi L_1 - \dots\right)(\varepsilon \boldsymbol{u}_1 + \varepsilon^2 \boldsymbol{u}_2 + \dots)$$
$$= \varepsilon^2 M_0 \boldsymbol{u}_1\boldsymbol{u}_1 + \varepsilon^3 (2M_0 \boldsymbol{u}_1\boldsymbol{u}_2 + N_0 \boldsymbol{u}_1\boldsymbol{u}_1\boldsymbol{u}_1) + O(\varepsilon^4) \, . \tag{2.4.4}$$

From the condition that this equation must hold to each order of ε, we obtain a set of balance equations in the form

$$\left(\frac{\partial}{\partial t} - L_0\right)\boldsymbol{u}_\nu = \tilde{\boldsymbol{B}}_\nu, \qquad \nu = 1, 2, \ldots, \tag{2.4.5}$$

where

$$\tilde{\boldsymbol{B}}_1 = 0, \tag{2.4.6a}$$

$$\tilde{\boldsymbol{B}}_2 = M_0\boldsymbol{u}_1\boldsymbol{u}_1, \tag{2.4.6b}$$

$$\tilde{\boldsymbol{B}}_3 = -\left(\frac{\partial}{\partial \tau} - \chi L_1 - D\nabla_s^2\right)\boldsymbol{u}_1 + 2M_0\boldsymbol{u}_1\boldsymbol{u}_2 + N_0\boldsymbol{u}_1\boldsymbol{u}_1\boldsymbol{u}_1, \tag{2.4.6c}$$

etc. The solvability condition is given by (2.2.13) with $\boldsymbol{B}_\nu$ replaced by $\tilde{\boldsymbol{B}}_\nu$. By decomposing $\tilde{\boldsymbol{B}}_\nu(t, \tau, \boldsymbol{s})$ into various harmonics,

$$\tilde{\boldsymbol{B}}_\nu(t, \tau, \boldsymbol{s}) = \sum_{l=-\infty}^{\infty} \tilde{\boldsymbol{B}}_\nu^{(l)}(\tau, \boldsymbol{s})\,\mathrm{e}^{\mathrm{i}l\omega_0 t},$$

the solvability condition may then be expressed as

$$\boldsymbol{U}^*\tilde{\boldsymbol{B}}_\nu^{(1)}(\tau, \boldsymbol{s}) = 0, \tag{2.4.7}$$

which is very similar to (2.2.14). The neutral solution is of the form

$$\boldsymbol{u}_1(t, \tau, \boldsymbol{s}) = W(\tau, \boldsymbol{s})\,\boldsymbol{U}\mathrm{e}^{\mathrm{i}\omega_0 t} + \text{c.c.} \tag{2.4.8}$$

It is straightforward to obtain $\boldsymbol{u}_2$ in terms of $\boldsymbol{u}_1$, and then $\tilde{\boldsymbol{B}}_3^{(1)}$ in terms of $\boldsymbol{u}_1$ (and hence of W). The result is

$$\tilde{\boldsymbol{B}}_3^{(1)} = -\left(\frac{\partial}{\partial \tau} - \chi L_1 - D\nabla_s^2\right)W\boldsymbol{U} + (2M_0\boldsymbol{U}\boldsymbol{V}_0 + 2M_0\bar{\boldsymbol{U}}\boldsymbol{V}_+ + 3N_0\boldsymbol{U}\boldsymbol{U}\bar{\boldsymbol{U}})\,|W|^2 W. \tag{2.4.9}$$

Finally, the solvability condition in (2.4.7) for $\nu = 3$ yields the Ginzburg-Landau equation

$$\frac{\partial W}{\partial \tau} = \chi\lambda_1 W + d\nabla_s^2 W - g|W|^2 W, \tag{2.4.10}$$

where d is generally a complex number given by

$$d \equiv d' + \mathrm{i}d'' = \boldsymbol{U}^*D\boldsymbol{U}, \tag{2.4.11}$$

while λ_1 and g are the same as before (2.2.5b, 20). The only difference between (2.4.10) and the Stuart-Landau equation is the existence of a diffusion term in (2.4.10). Note that if D is a scalar D_0, then d is real and equal to D_0. Generally, d' is positive. This is because the uniform steady solution is stable below criticality by assumption. In Appendix B, the derivation of the Ginzburg-Landau equation (in particular, the explicit calculation of λ_1, d, and g) is illustrated for a hypothetical reaction-diffusion model.

We want to see what the three asymptotic regimes (A) – (C) described in the previous section mean in the present context. The Ginzburg-Landau equation itself is considered to be valid for all these cases, but they show a clear contrast to one another in system size if measured in the new scale s. For simplicity, the system is supposed to have a comparable extension in all directions, and ξ measures its representative length. Since the representative system length $\tilde{\xi}$ in the new scale is given by $\tilde{\xi} = |\mu|^{1/2}\xi$ as is clear from the definition of s, we immediately have

$$\text{(A)}\ \tilde{\xi} = \infty, \quad \text{(B)}\ 0 < \tilde{\xi} < \infty, \quad \text{(C)}\ \tilde{\xi} = 0.$$

The physical meaning of the respective asymptotic regimes is now obvious. For case (C) the scaled size is zero, and the Ginzburg-Landau equation for a vanishing system size is nothing but the Stuart-Landau equation. For cases (A) and (B), the system leaves room for spatial variation, so that we have to work with the full Ginzburg-Landau equation. Studying cases (A) and (B) in turn may have its merits. Specifically, pattern formation in oscillating reaction-diffusion systems is considered to be a consequence of the interplay of a fairly large number of wavelike modes, and usually the effects of boundary conditions may be neglected. Then, it would be suitable to consider case (A), and this constitutes a part of the subject of Chap. 6. On the other hand, case (B) would be suited to the study of successive bifurcations and transitions to chaos. This is because these phenomena involve only a few effective degrees of freedom, which corresponds to a relatively small system size. Some forms of chemical turbulence will be considered from this viewpoint in Chap. 7.

It is sometimes more convenient to work with a further rescaled form of the Ginzburg-Landau equation. A suitable scale transformation would be

$$(\tau, s, W) \to (\sigma_1^{-1}\tau, \sqrt{d'/\sigma_1}s, \sqrt{\sigma_1/|g'|}W). \tag{2.4.12}$$

Since no confusion is expected, we shall express the scaled equation in terms of the more natural notations t and $\boldsymbol{r}$ in place of τ and $\boldsymbol{s}$. Then, under the condition (2.4.12), the Ginzburg-Landau equation above criticality reduces to the form

$$\frac{\partial W}{\partial t} = (1 + \mathrm{i}c_0)W + (1 + \mathrm{i}c_1)\nabla^2 W - (1 + \mathrm{i}c_2)|W|^2 W, \tag{2.4.13}$$

where

$$c_0 = \omega_1/\sigma_1, \quad c_1 = d''/d', \quad c_2 = g''/g', \tag{2.4.14}$$

and the bifurcation has been assumed to be supercritical (i.e., $g' > 0$). By further making the transformation $W \to W \exp(\mathrm{i} c_0 t)$, one may even eliminate c_0 to obtain

$$\frac{\partial W}{\partial t} = W + (1 + \mathrm{i} c_1) \nabla^2 W - (1 + \mathrm{i} c_2) |W|^2 W. \tag{2.4.15}$$

Thus, the only essential parameters are c_1 and c_2 apart from the system size. In later chapters, the form (2.4.14) or (2.4.15) will be used for preference.

When c_1 and c_2 are zero, the Ginzburg-Landau equation becomes identical to the small-amplitude equation related to some symmetry-breaking instabilities of the non-oscillatory type, such as the onset of Bénard convection and the appearance of Taylor vortices. On the contrary, it may happen that c_1 and c_2 become very large; a certain hypothetical reaction-diffusion model can in fact show this property (see Appendix B). In this limit, the Ginzburg-Landau equation reduces to the nonlinear Schrödinger equation, a well-known soliton-producing system appearing in many physical problems. An interesting fact which we find later is that chemical wavepatterns and chemical turbulence of a diffusion-induced type are possible only for regions intermediate between the two extremes.

Finally, we comment on some interrelationships between the Stuart-Landau or Ginzburg-Landau equation and a hypothetical mathematical model known as the λ-ω system with or without diffusion. The λ-ω system was introduced by Kopell and Howard (1973a) to demonstrate their general mathematical results concerning plane wave solutions of oscillatory reaction-diffusion systems. More recently, the same model was conveniently employed to investigate more complicated wave patterns such as circular waves (Greenberg, 1978) and rotating spiral waves (Yamada and Kuramoto, 1976a; Cohen et al., 1978; Greenberg, 1980). In the absence of diffusion, this model has the form

$$\frac{d}{dt}\begin{pmatrix} X \\ Y \end{pmatrix} = \begin{pmatrix} \lambda(R) & -\omega(R) \\ \omega(R) & \lambda(R) \end{pmatrix} \begin{pmatrix} X \\ Y \end{pmatrix}, \tag{2.4.16}$$

where $R = \sqrt{X^2 + Y^2}$ and $\lambda(R)$ and $\omega(R)$ are functions of R. It is clear that the Stuart-Landau equation becomes a special form of the λ-ω system by putting $X + \mathrm{i} Y = W$. If we interpret X and Y as chemical concentrations, the λ-ω system, as a hypothetical "chemical" system, may be generalized to include diffusion:

$$\frac{\partial}{\partial t}\begin{pmatrix} X \\ Y \end{pmatrix} = \begin{pmatrix} \lambda(R) + D_X \nabla^2 & -\omega(R) \\ \omega(R) & \lambda(R) + D_Y \nabla^2 \end{pmatrix} \begin{pmatrix} X \\ Y \end{pmatrix}, \tag{2.4.17}$$

$$D_X, D_Y \geq 0 .$$

It is important to realize that the Ginzburg-Landau equation is *not* a special case of (2.4.17). In fact, (2.4.13) may at best be written in the form

$$\frac{\partial}{\partial t}\begin{pmatrix} X \\ Y \end{pmatrix} = \begin{pmatrix} \lambda(R) + \nabla^2 & -\omega(R) - c_1 \nabla^2 \\ \omega(R) + c_1 \nabla^2 & \lambda(R) + \nabla^2 \end{pmatrix} \begin{pmatrix} X \\ Y \end{pmatrix}, \tag{2.4.18}$$

where

$$\lambda(R) = 1 - R^2, \qquad \omega(R) = c_0 - c_2 R^2 .$$

Equation (2.4.18) is not like the usual reaction-diffusion equations since the diffusion matrix has an antisymmetric part. This seemingly peculiar property is actually a general consequence of contracting the usual reaction-diffusion equations, for which the diffusion matrix may be a diagonal matrix of positive diffusion constants. On account of its sound physical basis, we shall use the Ginzburg-Landau equation in later chapters in preference to the λ-ω model. In particular, the existence of the c_1 terms will turn out to be crucial to the destabilization of uniform oscillations (see Appendix A) and hence to the occurrence of a certain type of chemical turbulence.

3. Method of Phase Description I

Weakly perturbed or weakly interacting finite-amplitude oscillations form a particular class of systems whose dynamics finds an extremely simplified description through the method presented here.

3.1 Systems of Weakly Coupled Oscillators

The neighborhood of the Hopf bifurcation point is itself an important asymptotic regime where the description of the dynamics is greatly simplified, as we saw in Chap. 2. Confining ourselves again to systems of oscillators which are coupled (but not necessarily through diffusion), there seems to exist at least one more physical situation for which an equally simplified description of the dynamics is expected. This is when the mutual coupling of the oscillators is weak; if necessary, weak external forces may be included. The aim of this chapter is to present a simple perturbation treatment appropriate for such circumstances.

What is meant by weak coupling may be clear if the oscillators are discrete, but caution is needed when they form a continuous field. For reaction-diffusion systems in particular, we mean that the diffusion terms are small compared to the reaction terms. Thus, if the composition vector $\boldsymbol{X}$ has some space dependence on an ordinary length scale or if $\nabla^2 \boldsymbol{X}$ is not small, then weak coupling means small diffusion constants. In contrast, if the diffusion constants are of ordinary magnitude, the same term means a slow space dependence of $\boldsymbol{X}$. Actually, however, no physical differences exist between these two interpretations; which view should be taken is merely a matter of convenience or taste. For definiteness, we shall take the latter view throughout, since this is also in accord with the discussion in Chap. 2, where we thought that the system size might become very large, while the diffusion constants were assumed to be of ordinary magnitude. In this chapter we do not try to find conditions for which assuming a slow space dependence of $\boldsymbol{X}$ is justified; this forms a separate problem, and will be considered in Chaps. 6 and 7. Perturbation theories may be developed by simply assuming a slow space dependence of $\boldsymbol{X}$ without inquiring into why that is.

Some peculiar features of the perturbation theory given here seem worth mentioning. The only assumption to be made in our theory is that the individual oscillators are weakly perturbed, and nothing is assumed about the specific nature of those oscillators. They may be quite general ones, except that they obey a system of n ordinary differential equations; the system need not lie near some bifurcation point nor in any other extreme situation. Since limit cycle motion cannot be solved analytically in general, this means that even the reference motion (i.e. unperturbed motion) cannot be expressed analytically. In this respect, the present theory differs considerably from the reductive perturbation method of Chap. 2 and also from ordinary small-parameter methods or quasi-linear theories of nonlinear oscillations such as developed in the book by Bogoliubov and Mitropolsky (1961). It may be wondered what the use of this kind of peculiar perturbation theory is, and we now explain briefly. It is true that the motion of the natural oscillators cannot be represented analytically. But we know at least that they make strictly periodic oscillations as $t \to \infty$. One may then introduce a phase ϕ into the limit cycle orbit of each oscillator in such a way that the periodic motion on it may produce a constant increase of ϕ, e.g., $d\phi/dt = 1$. Suppose some weak influences from the outside world (e.g., from the remaining oscillators) have been switched on. Each limit cycle is supposed to possess more or less "stiffness" in orbital shape against perturbations, so that when only weakly perturbed, the state point of a given oscillator hardly deviates from its natural closed orbit but remains almost on it. But the deviation in phase produced along the natural orbit could accumulate, thus needing a proper description. Since this argument applies to any oscillator, we are led to the following picture: the state of each oscillator can be approximately specified by its phase value, and its rate of change is determined by the phase values of all the other oscillators interacting with it. Thus the dynamics of our system of N discrete oscillators may be reduced to N coupled ordinary differential equations for N phase variables. For diffusion-coupled continua, a partial differential equation for a space-time dependent phase may be obtained in a closed form. In this way, the assumption of weak coupling leads to a phase description, thereby achieving a considerable reduction of the degrees of freedom.

A further contraction of the dynamics is made possible by virtue of the same starting assumption. This comes from the fact that weak perturbations generally produce a long time scale in the dynamics compared to the period of the natural oscillations. This time scale characterizes the slow evolution of the phase disturbances due to the perturbation. Such a clear separation in time scale enables us to average rapidly oscillating quantities appearing in the evolution equation for slow variables. Even after such simplifications have been made, the equations for phases may still retain some unknown quantities which depend, e.g., on the orbital shape of the natural oscillators, the specific form of the mutual coupling, etc. This is a price we have to pay for the fact that the unperturbed system is itself a general, analytically unsolvable system. Fortunately, such unknown quantities often appear merely as a few numerical coefficients or at worst one or two periodic functions of the phases. Without knowing these, one may still analyze the equations to a considerable extent and draw many useful conclusions. If necessary, one may even calculate such unknowns separately, e.g., with the aid of a computer, for a given model. The final question may be what is the use of knowing the phase states of the individual oscillators, since phase is, after all, the quantity we defined in a rather arbitrary way. In answer, we would say that the phase provides important information independently of the way it is defined. For instance, whether a given pair of oscillators are mutually entrained to an identical frequency may be found out from the phase difference between the two as a function of t; if it is bounded, they are entrained, and if not, non-entrained. The utility of the phase description may also be realized from the following observation. If our assumption of small orbital deformation is valid, the basic structure of various wave patterns in oscillating reaction-diffusion fields may be represented by the contours of constant phase because they approximate equi-concentration contours. For some turbulence problems, too, the phase description may prove useful because if the evolution of phases turns out to be chaotic (non-chaotic), one may conclude that the original dynamics is also chaotic (non-chaotic).

An attempt to describe populations of oscillators in terms of phases was made by Winfree (1967), although the theory involved some drastic assumptions. More recently, Neu (1979b, 1980) developed a phase description method for discrete populations. In the context of reaction-diffusion dynamics, Ortoleva and Ross (1973, 1974) were the first to derive a partial differential equation for the phase in the discussion of phase waves. However, an important nonlinear term representing the effect of frequency modification due to a phase gradient was lacking in their phase diffusion equation, and this was properly taken into account later (Kuramoto and Tsuzuki, 1976; Kuramoto and Yamada, 1976b; Neu, 1979a). The phase description method presented below seems to be the simplest of all the related theories so far. Although the method proves useful enough to treat mutual synchronization in populations (Chap. 5) and some wave phenomena (Chap. 6), the potential power of the phase description idea cannot fully find its expression in the formulation of this chapter. Its more elaborate formulation which enables a considerable expansion of its applicability will form the subject of Chap. 4.

3.2 One-Oscillator Problem

Although the system with which we are ultimately concerned comprises the continuous fields or discrete populations of coupled oscillators, it seems appropriate to begin with a simpler system whose study fully illustrates the perturbation idea. Once the method has been formulated for it, its extension to more complicated systems will turn out to be extremely easy.

Let $X_0(t)$ denote a linearly stable T-periodic solution of an n-dimensional system of ordinary differential equations (2.1.1), or

$$\frac{dX_0}{dt} = F(X_0) , \quad X_0(t+T) = X_0(t) . \tag{3.2.1}$$

Let the vector field be perturbed as

$$\frac{dX}{dt} = F(X) + \varepsilon p(X) , \tag{3.2.2}$$

where $\varepsilon p(X)$ represents a small perturbation generally depending on X, and ε is an indicator of the smallness of p and is finally set equal to 1. We shall, however, sometimes use expressions like "terms of order ε^2", which should be understood as "terms of order $|p|^2$". Periodic motions will persist when the perturbation εp is introduced, but its period will deviate slightly from T. We now want to find a formal expression for this change in period to the lowest order in ε. Although this might look like a problem of little practical interest, one may easily generalize the class of perturbations εp, and then the theory proves to cover a variety of phenomena of practical interest.

Let C denote the closed orbit corresponding to $X_0(t)$. Since C is supposed to be stable, each state point X in the vicinity of C approaches C as $t \to \infty$ in the absence of perturbation. In describing this asymptotic periodic motion on C, we associate a certain value of a scalar ϕ to each $X \in C$ in such a way that the motion on C may produce a constant increase of ϕ, or specifically,

$$\frac{d\phi(X)}{dt} = 1\,, \quad X \in C\,. \tag{3.2.3}$$

The quantity ϕ may be called the phase defined on C, and its value is only determined to an integer multiple of T. It would be very inconvenient, however, if the definition of phase were restricted to C. This is because arbitrarily small perturbations could generally kick the state point out of C, so that without definition of ϕ outside C we could no longer say anything about the phase of *perturbed* oscillations. It is therefore desirable to extend the definition of ϕ so that we could say, e.g., how the phase, or its rate of change, is influenced by the perturbation. Since the perturbation is assumed to be weak, ϕ need only be defined in the vicinity of C.

To make the picture clearer, we imagine a circular tube through which the orbit C threads (Fig. 3.1). We want to define $\phi(X)$ for each X inside the tube. We use here a language appropriate to a three-dimensional state space, but actually we are working with an $n(\geq 2)$-dimensional system. Let G denote this n-dimensional tubular region containing all neighborhoods of C. The domain of attraction of C is assumed to contain G inside it. The tube may be thin to the extent that the perturbation is weak.

One may adopt the following definition of $\phi(X)$, which seems to be most natural and convenient. Our definition is essentially the same as what is called "asymptotic phase" (Coddington and Levinson, 1955). Let P denote a point

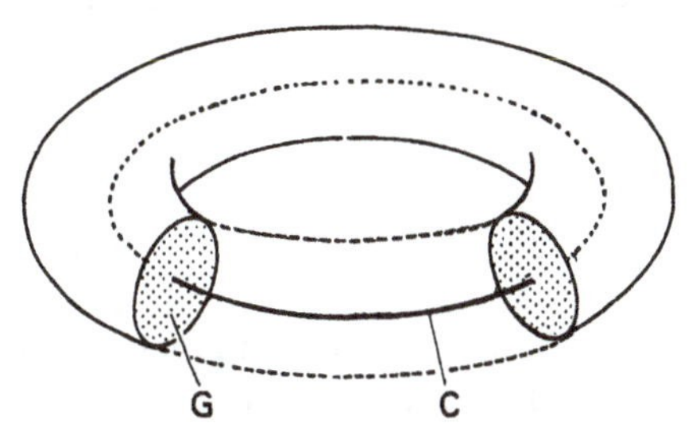

Fig. 3.1. Limit cycle orbit enclosed in a thin tube

such that $P \in G$ but $P \notin C$. Let Q denote another point lying on C. One may apply the definition of ϕ on C, and associate some phase value ϕ_Q to Q. We now let P and Q start to move simultaneously according to the unperturbed equation of motion (2.1.1). Then P will approach C, but as $t \to \infty$, P and Q will generally be separated by a finite distance from each other on C. Then we say that ϕ_P, i.e., the initial phase of P, differs from ϕ_Q, i.e., the initial phase of Q. It may happen that P and Q come infinitely close to each other as $t \to \infty$. Then we say that $\phi_P = \phi_Q$. Since P may be an arbitrary point in G, this shows one possible way of associating a certain phase value to each point in G. One may imagine that G is completely filled with a one-parameter family of hypersurfaces of constant phase, each of which is $(n-1)$-dimensional. The hypersurfaces thus defined, which are denoted by $I(\phi)$, have a close connection with the notion of "isochrons" which was first introduced by Winfree (1967) for practical purposes and later investigated mathematically by Guckenheimer (1975); the original notion of isochrons is considered to be an extension of $I(\phi)$ to the entire state space, and it therefore generated some topologically interesting problems in connection with *phaseless sets*. It immediately follows from our definition of ϕ that if a given set of state points in G happen to find themselves on a common isochron at a certain time, they always stay on a common isochron as far as they remain unperturbed. In particular, the rate of change of ϕ of a given state point is the same, whether it belongs to C or not. Thus,

$$\frac{d\phi(X)}{dt} = 1\,, \quad X \in G\,. \tag{3.2.4}$$

On the other hand, we have the obvious identity

$$\frac{d\phi(X)}{dt} = \mathrm{grad}_X\phi \cdot \frac{dX}{dt}\,, \tag{3.2.5}$$

so that (3.2.4) combined with (2.1.1) gives

$$\mathrm{grad}_X\phi \cdot F(X) = 1\,, \quad X \in G\,. \tag{3.2.6}$$

It is also clear that if a stroboscopic observation of the unperturbed system is made at times $t = nT$ $(n = 0, 1, 2, \ldots)$, then the state point will always be found on an identical isochron and closer and closer to C with n.

Having thus defined the phase for any X near C, it now makes sense to say something about perturbed motions of ϕ. In fact, substituting (3.2.2) into (3.2.5) and using (3.2.6), we obtain

$$\begin{aligned} \frac{d\phi(X)}{dt} &= \mathrm{grad}_X\phi \cdot [F(X) + \varepsilon p(X)] \\ &= 1 + \varepsilon\, \mathrm{grad}_X\phi \cdot p(X)\,. \end{aligned} \tag{3.2.7}$$

This equation is still exact. It is seen that the right-hand side depends generally on the precise location of point X on the same $I(\phi)$; that is, the evolution equation

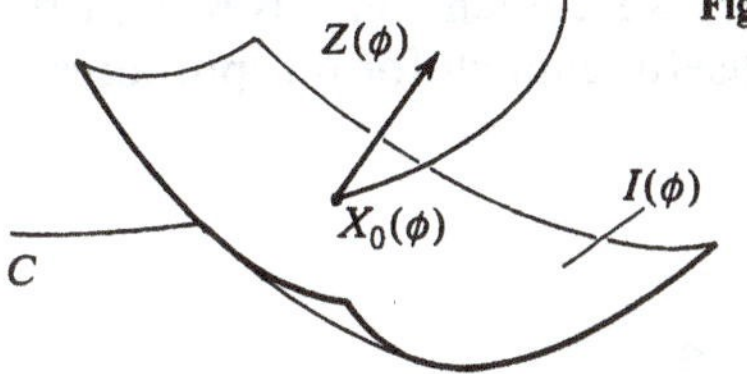

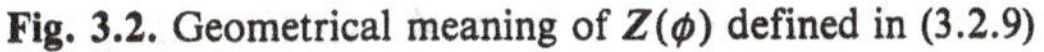

Fig. 3.2. Geometrical meaning of $Z(\phi)$ defined in (3.2.9)

for ϕ does not have a closed form in ϕ. What is important here is that a perturbation idea can make it closed. The reason is the following: although specifying the value of ϕ is insufficient for locating the point X on $I(\phi)$, we know at least that X is close to $X_0(\phi)$ [i.e., the intersection point of $I(\phi)$ and C]; the deviation $|X - X_0(\phi)|$ is considered to go to zero as $\varepsilon \to 0$, as far as its asymptotic behavior as $t \to \infty$ is concerned. Thus, one is permitted to replace X with $X_0(\phi)$ on the right-hand side of (3.2.7) in the lowest-order approximation. This leads to a self-contained equation for ϕ:

$$\frac{d\phi}{dt} = 1 + \varepsilon\Omega(\phi)\,, \qquad \text{where} \tag{3.2.8}$$

$$\begin{aligned} \Omega(\phi) &= \boldsymbol{Z}(\phi) \cdot \boldsymbol{\Pi}(\phi)\,, \\ \boldsymbol{Z}(\phi) &= (\operatorname{grad}_X \phi)_{X = X_0(\phi)}\,, \\ \boldsymbol{\Pi}(\phi) &= \boldsymbol{p}(X_0(\phi))\,. \end{aligned} \tag{3.2.9}$$

The vector $\boldsymbol{Z}(\phi)$ may be called the (phase-dependent) sensitivity (Winfree, 1967), as it measures how sensitively the oscillator responds to external perturbations. Figure 3.2 shows a geometrical interaction of $\boldsymbol{Z}(\phi)$. It is represented by a vector based at the point $X_0(\phi)$ and normal to $I(\phi)$, its length being given by the number density of surfaces I at $X_0(\phi)$. Note that $\boldsymbol{Z}(\phi)$ and $\boldsymbol{\Pi}(\phi)$ are T-periodic functions of ϕ, which means that the instantaneous frequency, i.e., the right-hand side of (3.2.8), is T-periodic in ϕ.

In order to obtain the average frequency, we introduce the phase disturbance ψ via $\phi = t + \psi$, and express (3.2.8) in the form

$$\frac{d\psi}{dt} = \varepsilon\Omega(t + \psi)\,. \tag{3.2.10}$$

This equation shows that ψ is a slow variable, so that it hardly changes during the period T. Then, (3.2.10) may be time averaged as

$$\frac{d\psi}{dt} = \varepsilon\omega\,, \qquad \omega = \frac{1}{T}\int_0^T \Omega(t)\,dt\,, \tag{3.2.11}$$

which gives the frequency change we wanted to obtain. Such an averaging procedure may look a little ambiguous. A more systematic treatment exists and will be developed in Chap. 4. In this chapter, we content ourselves with the simple method above which in fact turns out useful enough for our purposes in Chaps. 5 and 6.

3.3 Nonlinear Phase Diffusion Equation

One may wonder how the method described above can be extended to many-oscillator systems. In this section we will consider reaction-diffusion systems for which the above treatment is most easily generalized.

Reaction-diffusion equations may be written in the form of (3.2.2) if $\boldsymbol{p}$ is interpreted as a Laplacian operator multiplied by the matrix D:

$$\varepsilon \boldsymbol{p} = D\nabla^2 . \tag{3.3.1}$$

When we take the diffusion to be a small perturbation, it is presupposed that the spatial variation of $\boldsymbol{X}$ is slow, the typical wavelength being of order $\varepsilon^{-1/2}$. It is easy to confirm that the lowest order perturbation results (3.2.8, 9) do not change at all when $\boldsymbol{p}$ is changed to a differential operator. Thus, applying (3.3.1) to (3.2.9), and setting ε to 1, we get a partial differential equation:

$$\frac{\partial\phi}{\partial t} = 1 + \Omega^{(1)}(\phi)\nabla^2\phi + \Omega^{(2)}(\phi)(\nabla\phi)^2 , \tag{3.3.2}$$

where

$$\begin{aligned} \Omega^{(1)}(\phi) &= \boldsymbol{Z}(\phi)D\frac{d\boldsymbol{X}_0(\phi)}{d\phi} , \\ \Omega^{(2)}(\phi) &= \boldsymbol{Z}(\phi)D\frac{d^2\boldsymbol{X}_0(\phi)}{d\phi^2} . \end{aligned} \tag{3.3.3}$$

Note that $\Omega^{(1)}$ and $\Omega^{(2)}$ are T-periodic in ϕ. In terms of ψ, (3.3.2) is transformed to

$$\frac{\partial\psi}{\partial t} = \Omega^{(1)}(t+\psi)\nabla^2\psi + \Omega^{(2)}(t+\psi)(\nabla\psi)^2 . \tag{3.3.4}$$

After averaging the periodic coefficients over the period T, which is justified for the same reason as in Sect. 3.2, we have

$$\frac{\partial\psi}{\partial t} = \alpha\nabla^2\psi + \beta(\nabla\psi)^2 , \quad \text{or} \tag{3.3.5}$$

$$\frac{\partial\phi}{\partial t} = 1 + \alpha\nabla^2\phi + \beta(\nabla\phi)^2 , \quad \text{where} \tag{3.3.6}$$

$$\alpha = \frac{1}{T}\int_0^T \Omega^{(1)}(t)\,dt\,, \quad \beta = \frac{1}{T}\int_0^T \Omega^{(2)}(t)\,dt\,. \tag{3.3.7}$$

Equation (3.3.5) represents a nonlinear phase diffusion equation. It is equivalent to the Burgers equation in the case of one space dimension (Chap. 6). It is known that the Burgers equation can be reduced to a linear diffusion equation through a transformation called the Hopf-Cole transformation (Burgers, 1974), and essentially the same is true for (3.3.5) in an arbitrary dimension. We shall take advantage of this fact in Chap. 6 when analytically discussing a certain form of chemical waves.

3.4 Representation by the Floquet Eigenvectors

The notion of asymptotic phase or isochrons has a certain relationship to the eigenvectors, defined for the linearized system of (2.1.1) about its periodic solution $X_0(t)$. Consequently, various quantities which appear in our perturbation results, e.g., α and β, may be reinterpreted in terms of those eigenvectors. This new interpretation may sometimes prove convenient for analytical or numerical calculations based on a specific model (Sect. 3.5). Moreover, such a reformulation opens the way to a more systematic method of phase description as developed in Chap. 4.

There is a theory called the Floquet theory (Cesari, 1971) which concerns first-order linear systems with periodic coefficients. In the present context, such a system arises from the linearization of (2.1.1) about $X_0(t)$. By putting $X(t) = X_0(t) + u(t)$, this leads to

$$\frac{du}{dt} = L(t)u\,, \tag{3.4.1}$$

where L is an $n \times n$ T-periodic matrix with its ijth element given by $L_{ij} = \partial F_i(X_0(t))/\partial X_{0j}(t)$. According to the Floquet theory, it is possible to define an eigenvalue problem for (3.4.1). This is based on the fact that the general solution of (3.4.1) is expressed in the form

$$u(t) = S(t)\,e^{\Lambda t}u(0)\,. \tag{3.4.2}$$

Here $S(t)$ is a T-periodic matrix with the initial condition $S(0) = 1$, and Λ is some time-independent matrix. In the next chapter, the identity

$$\frac{dS(t)}{dt} + S(t)\Lambda - L(t)S(t) = 0\,, \tag{3.4.3}$$

which follows from (3.4.1, 2), will frequently be used. Let u_l and u_l^* denote the right and left eigenvectors of Λ, the corresponding eigenvalues being denoted by λ_l, i.e.,

$$\Lambda u_l = \lambda_l u_l, \quad u_l^* \Lambda = \lambda_l u_l^*, \quad l = 0, 1, \ldots, n-1.$$

The eigenvalues are assumed to be algebraically simple, and one may require the orthonormality condition

$$u_l^* \cdot u_m = \delta_{lm}, \quad l, m = 0, 1, \ldots, n-1.$$

Since $X_0(t)$ is assumed stable, no eigenvalues have a positive real part. It is also clear that autonomous oscillations generally have one special eigenvalue which is identically zero. This corresponds to phase disturbances, i.e., disturbances produced along the closed orbit. Let λ_0 denote the zero eigenvalue, and assume that the remaining $n-1$ eigenvalues have negative real parts. The zero eigenvector u_0 may be taken as

$$u_0 = \left(\frac{dX_0(t)}{dt}\right)_{t=0}, \tag{3.4.4}$$

because the right-hand side gives a tangent vector to C at point $X_0(0)$ and hence has the same direction as that of the infinitesimal phase disturbances. More generally, one may show that

$$S(t)u_0 = \frac{dX_0(t)}{dt}. \tag{3.4.5}$$

In fact, by differentiating (3.2.1), we have

$$\frac{d}{dt}\left(\frac{dX_0(t)}{dt}\right) = L(t)\left(\frac{dX_0(t)}{dt}\right),$$

which shows that $dX_0(t)/dt$ is a particular solution of (3.4.1). Then, according to (3.4.2), $dX_0(t)/dt$ must be of the form

$$\frac{dX_0(t)}{dt} = S(t)\,\mathrm{e}^{\Lambda t}\left(\frac{dX_0(t)}{dt}\right)_{t=0}, \tag{3.4.6}$$

which reduces to (3.4.5) via (3.4.4) and the fact that $\Lambda u_0 = 0$.

Let $T(\phi)$ represent the $(n-1)$-dimensional hyperplane tangent to the isochron $I(\phi)$ at point $X_0(\phi)$ (Fig. 3.3). We want to show that $T(0)$ forms the eigenspace spanned by the $n-1$ eigenvectors $u_l (l \neq 0)$. Let us first note the following: for system (2.1.1), imagine two state points $X_0(t) \in C$ and $X(t)$, both lying on the same I initially (and hence for any $t \geq 0$). Then the relative position vector $u(t)$ $[=X(t) - X_0(t)]$ shrinks to zero as t goes to infinity. If $|u(t)|$ is sufficiently small,

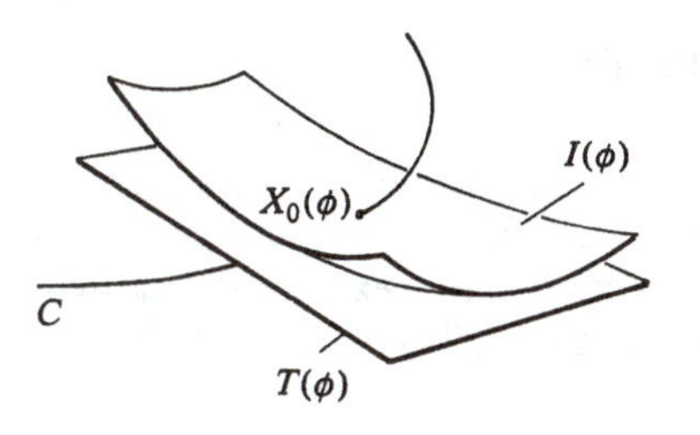

Fig. 3.3. $(n-1)$-dimensional hyperplane $T(\phi)$ tangent to the isochron $I(\phi)$ at point $X_0(\phi)$ lying on the limit cycle orbit C

then in the statement above one may replace I by its tangent space T. That is, any vector $\boldsymbol{u}(t)$ lying initially on $T(0)$ has the property that $\boldsymbol{u}(t)\rightarrow 0$ as $t\rightarrow\infty$ for the linearized system (3.4.1). Furthermore, (3.4.2) implies that this behavior of $\boldsymbol{u}(t)$ is possible only if $\boldsymbol{u}(0)$ is free from the zero-eigenvector component, which is nothing but the fact we wanted to show. This fact, combined with the property that $\boldsymbol{Z}(\phi)$ is normal to $T(\phi)$, leads to the relation

$$\boldsymbol{Z}(0)\cdot\boldsymbol{u}_l=0\,,\qquad l\neq 0\,,$$

which says that $\boldsymbol{Z}(0)$ is proportional to $\boldsymbol{u}_0^*$; the proportionality constant may be taken to be 1, or

$$\boldsymbol{u}_0^*=\boldsymbol{Z}(0)\,, \tag{3.4.7}$$

because the above choice for the length of $\boldsymbol{u}^*$ satisfies the normalization condition $\boldsymbol{u}_0^*\cdot\boldsymbol{u}_0=1$ or $\boldsymbol{Z}(0)\cdot(d\boldsymbol{X}_0(t)/dt)_{t=0}=1$. In fact, a more general equality

$$\boldsymbol{Z}(t)\cdot\frac{d\boldsymbol{X}_0(t)}{dt}=1 \tag{3.4.8}$$

follows from (3.2.6) by choosing point $\boldsymbol{X}$ on C. Finally, we note that (3.4.5) and (3.4.8) are compatible only if

$$\boldsymbol{Z}(t)=\boldsymbol{u}_0^*S^{-1}(t)\,. \tag{3.4.9}$$

With the use of (3.4.5, 9), $\Omega(\phi)$, $\Omega^{(1)}(\phi)$, and $\Omega^{(2)}(\phi)$ are reexpressed as

$$\Omega(\phi)=\boldsymbol{u}_0^*S^{-1}(\phi)\Pi(\phi)\,, \tag{3.4.10a}$$

$$\Omega^{(1)}(\phi)=\boldsymbol{u}_0^*S^{-1}(\phi)DS(\phi)\boldsymbol{u}_0\,, \tag{3.4.10b}$$

$$\Omega^{(2)}(\phi)=\boldsymbol{u}_0^*S^{-1}(\phi)D\frac{dS(\phi)}{d\phi}\boldsymbol{u}_0\,. \tag{3.4.10c}$$

3.5 Case of the Ginzburg-Landau Equation

It would be instructive here to illustrate the theory presented above with a simple reaction-diffusion model. A suitable model would be the Ginzburg-Landau equation in the form of (2.4.13). As noted in Sect. 2.4, it is expressed as a two-component reaction-diffusion system, although the diffusion matrix D then involves an antisymmetric part:

$$D = \begin{pmatrix} 1 & -c_1 \\ c_1 & 1 \end{pmatrix}, \tag{3.5.1}$$

unlike diffusion matrices in ordinary reaction-diffusion systems. We want to calculate $\Omega^{(1)}(\phi)$ and $\Omega^{(2)}(\phi)$ for this system via (3.4.10b, c). For this purpose, we need to calculate $\boldsymbol{u}_0$, $\boldsymbol{u}_0^*$, and $S(\phi)$. Since these are quantities all associated with diffusionless systems, we need only consider the Stuart-Landau equation

$$\frac{dW}{dt} = (1+\mathrm{i}c_0)W - (1+\mathrm{i}c_2)|W|^2 W, \quad \text{or} \tag{3.5.2}$$

$$\begin{aligned} \frac{dX}{dt} &= X - c_0 Y - (X - c_2 Y)(X^2 + Y^2), \\ \frac{dY}{dt} &= Y + c_0 X - (Y + c_2 X)(X^2 + Y^2). \end{aligned} \tag{3.5.3}$$

Let $w(t)$ denote a disturbance variable defined by

$$W(t) = W_0(t)[1 + w(t)], \tag{3.5.4}$$

where $W_0(t)$ is the periodic solution of (3.5.2), or

$$\begin{aligned} &W_0(t) \equiv X_0(t) + \mathrm{i}Y_0(t) = \exp(\mathrm{i}\omega_0 t), \\ &\quad \omega_0 = c_0 - c_2. \end{aligned} \tag{3.5.5}$$

The linearization of (3.5.2) about $W_0(t)$ gives

$$\frac{dw}{dt} = -(1+\mathrm{i}c_2)(w + \bar{w}). \tag{3.5.6}$$

If we put $w = \xi + \mathrm{i}\eta$, (3.5.6) may be expressed as

$$\frac{d}{dt}\begin{pmatrix} \xi \\ \eta \end{pmatrix} = \Lambda \begin{pmatrix} \xi \\ \eta \end{pmatrix},$$

or by integration,

$$\begin{pmatrix} \xi(t) \\ \eta(t) \end{pmatrix} = \mathrm{e}^{\Lambda t} \begin{pmatrix} \xi(0) \\ \eta(0) \end{pmatrix}, \tag{3.5.7}$$

where

$$\Lambda = -2 \begin{pmatrix} 1 & 0 \\ c_2 & 0 \end{pmatrix}. \tag{3.5.8}$$

In order to obtain $\boldsymbol{u}_0$, $\boldsymbol{u}_0^*$, and $S(\phi)$, we have to linearize (3.5.3) in the disturbances $(x,y) \equiv (X - X_0(t), Y - Y_0(t))$ and find their solution in the form of (3.4.2). This is easy to do if we apply the relation between (x,y) and (ξ,η), which is given by (3.5.4), to (3.5.7). One may write (3.5.4) more explicitly as

$$\begin{pmatrix} x \\ y \end{pmatrix} = S(t) \begin{pmatrix} \xi \\ \eta \end{pmatrix}, \tag{3.5.9}$$

where

$$S(t) = \begin{pmatrix} \cos\omega_0 t & -\sin\omega_0 t \\ \sin\omega_0 t & \cos\omega_0 t \end{pmatrix}. \tag{3.5.10}$$

Thus (3.5.7) is equivalent to

$$\begin{pmatrix} x(t) \\ y(t) \end{pmatrix} = S(t)\mathrm{e}^{\Lambda t} \begin{pmatrix} x(0) \\ y(0) \end{pmatrix} \tag{3.5.11}$$

which is certainly of the same form as (3.4.2). Calculation of the eigenvectors of Λ is straightforward. We obtain

$$\boldsymbol{u}_0 = \omega_0 \begin{pmatrix} 0 \\ 1 \end{pmatrix}, \tag{3.5.12a}$$

where the factor ω_0 is needed for consistency with (3.4.4). We further get

$$\boldsymbol{u}_0^* = \omega_0^{-1}(-c_2, 1), \tag{3.5.12b}$$

$$\boldsymbol{u}_1 = \begin{pmatrix} 1 \\ c_2 \end{pmatrix}, \tag{3.5.12c}$$

$$\boldsymbol{u}_1^* = (1, 0). \tag{3.5.12d}$$

The eigenvalues are $\lambda_0 = 0$ and $\lambda_1 = -2$. By applying the above values of S, $\boldsymbol{u}_0$, and $\boldsymbol{u}_0^*$ to (3.4.10b, c), we find that $\Omega^{(1)}$ and $\Omega^{(2)}$ are ϕ-independent and given by

$$\Omega^{(1)} = \alpha = 1 + c_1 c_2 , \tag{3.5.13a}$$

$$\Omega^{(2)} = \beta = \omega_0 (c_2 - c_1) . \tag{3.5.13b}$$

The nonlinear phase diffusion equation (3.3.5) now takes the explicit form

$$\frac{\partial \psi}{\partial t} = (1 + c_1 c_2) \nabla^2 \psi + \omega_0 (c_2 - c_1)(\nabla \psi)^2 . \tag{3.5.14}$$

For the unperturbed orbit, $d\psi/dt = 0$ and $d\Theta/dt = \omega_0$, which implies the relation $\theta = \omega_0(t + \psi)$ for a weak perturbation, where θ is the phase of W. It follows, therefore, that whenever the orbital deformation due to the diffusion coupling is negligible, the Ginzburg-Landau equation is contracted to

$$\begin{aligned} &R = 1 , \\ &\frac{\partial \Theta}{\partial t} = \omega_0 + (1 + c_1 c_2) \nabla^2 \Theta + (c_2 - c_1)(\nabla \Theta)^2 . \end{aligned} \tag{3.5.15}$$

This reduced form breaks down if $1 + c_1 c_2 < 0$, which actually occurs, at least for a certain hypothetical chemical reaction model as shown in Appendix B. A negative phase diffusion coefficient implies turbulence, and this problem will be revisited in Chap. 7.

4. Method of Phase Description II

A deeper meaning of the phase description method of Chap. 3 manifests itself as we cast it into a more systematic formulation. This will widen the scope of the method, thereby encompassing problems not necessarily inherent in oscillator systems.

4.1 Systematic Perturbation Expansion

The first method of phase description (abbreviated as Method I hereafter) described in Chap. 3 has the advantage of being extremely simple and capable of dealing with various types of perturbations as the $\varepsilon \boldsymbol{p}$ term. Such an advantage will fully be utilized in later chapters, especially in Chap. 5. It is also true, however, that Method I has a strong limitation in that it only serves as a lowest-order perturbation theory. There exist at least two practical reasons why Method I should be reformulated and generalized into a systematic perturbation expansion. The first reason is that some physically important situations exist where the lowest-order theory breaks down while the original perturbation idea itself remains valid. This occurs when the phase diffusion coefficient α becomes slightly negative in (3.3.5) so that the system shows unphysical behavior if higher-order terms are not included. Which terms should be included, or how to analyze the modified equation obtained in this way will be considered in Chap. 7. In the present chapter, we are only concerned with a formal perturbation expansion. The second reason is the following: by associating with ϕ some physical quantities different from the phase of oscillation, the phase description idea, if formulated systematically, turns out to encompass new classes of phenomena which may even be irrelevant to oscillatory dynamics. The fact that the system consists of oscillating units is then no longer a vital condition to the applicability of the theory. There seems to exist a more basic property which the systems must commonly possess for our extended perturbation idea to work. This is the property that the eigenvalue spectrum of the system linearized about a certain reference state has a phaselike branch which extends from the zero eigenvalue. Some examples of such systems will be treated in Sects. 4.3 and 4.

This section is devoted to a reconsideration of system (3.2.2), to obtain the frequency change as a power series in ε. A slightly careful examination of Method I reveals that a major obstacle to systematic perturbation expansions lies in the fact that the surfaces of constant phase are generally curved in state space. Since the definition of the surfaces of constant phase is entirely at our disposal,

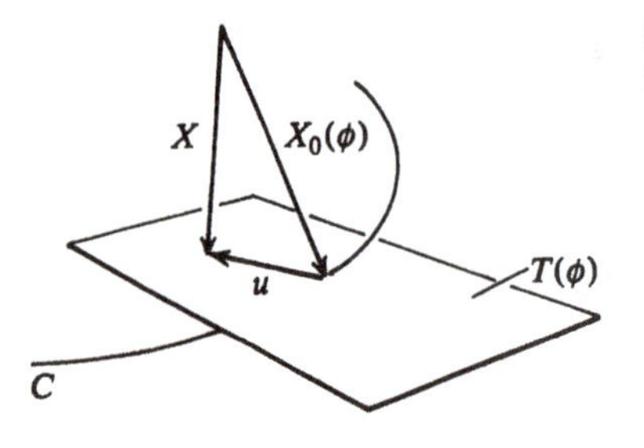

Fig. 4.1. Splitting of the state vector X into the unperturbed part $X_0(\phi)$ and the deviation u

we should redefine them so as to meet our present purposes. A convenient choice would be such that the hypersurface of some phase value ϕ be identified with $T(\phi)$ rather than $I(\phi)$. We shall adopt this definition below. Different $T(\phi)$ may intersect in a messy way, which might seem to cause trouble. However, this can only happen at a finite distance from the closed orbit C, so that no trouble actually arises in developing those perturbation theories which are based asymptotic expansions about C.

Imagine a tubular region G as in Sect. 3.2 (Fig. 3.1). It is supposed that G is so thin that no intersections between different $T(\phi)$ occur within G. Assume further that the perturbation εp is sufficiently weak so that no state points can get out of G for all t possibly except for initial transients. Imagine a state point $X(t)$ which obeys the perturbed equation (3.2.2). It will move from one T to another T, and we choose $\phi(t)$ in such a way that $X(t)$ is found on $T(\phi(t))$ for any t. Note that $\phi(t)$ represents the phase of $X(t)$ in the new definition of phase. For any t, $X(t)$ can then be split uniquely as

$$X(t) = X_0(\phi) + u(t) \tag{4.1.1}$$

(Fig. 4.1). Since the state points $X(t)$ and $X_0(\phi)$ belong simultaneously to $T(\phi)$, the deviation vector $u(t)$ should lie on $T(\phi)$. This means that

$$Z(\phi) \cdot u(t) = 0 \,, \tag{4.1.2}$$

which represents a convenient property resulting from the new definition of phase; if we worked with the old definition, the validity of (4.1.2) would be restricted to the linear regime in u.

The evolution of $X(t)$ may now be completely specified by specifying $u(t)$ and $\phi(t)$. In place of $\phi(t)$, one may alternatively take the quantity $\Omega(t)$, i.e., the deviation in the instantaneous frequency defined by

$$\frac{d\phi}{dt} = 1 + \Omega(t) \,. \tag{4.1.3}$$

If no perturbations are present, we have

$$u(t) \,, \Omega(t) \to 0 \quad \text{as} \quad t \to \infty \,,$$

while in the presence of the perturbation $\varepsilon \boldsymbol{p}$, we expect that

$$\boldsymbol{u}(t), \Omega(t) \to O(\varepsilon) \quad \text{as} \quad t \to \infty .$$

In contrast to $\Omega(t)$, the phase deviation $\phi(t)-t$ is generally unbounded as a function of t when $\varepsilon \neq 0$, so that the latter is not an appropriate quantity to treat perturbatively. Essentially the same idea underlies all perturbation theories of nonlinear oscillations involving the elimination of secular effects. Below, we treat $\boldsymbol{u}(t)$ and $\Omega(t)$ perturbatively.

The next essential step is to adopt the view that $\boldsymbol{u}(t)$ and $\Omega(t)$ are dependent on t through ϕ. To express this clearly, we use the notation $\boldsymbol{u}(\phi)$, $\Omega(\phi)$, and $\boldsymbol{X}(\phi)$ for $\boldsymbol{u}(t)$, $\Omega(t)$, and $\boldsymbol{X}(t)$, respectively. Thus (4.1.1) should be expressed as

$$\boldsymbol{X}(\phi) = \boldsymbol{X}_0(\phi) + \boldsymbol{u}(\phi) , \tag{4.1.4}$$

where $\boldsymbol{u}$ must satisfy (4.1.2) or

$$\boldsymbol{u}_0^* S^{-1}(\phi)\boldsymbol{u}(\phi) = 0 . \tag{4.1.5}$$

Further,

$$\frac{d\phi}{dt} = 1 + \Omega(\phi) , \tag{4.1.6}$$

which is nothing but a nonlinear evolution equation for ϕ in a self-contained form.

Equation (3.2.2) now becomes

$$\frac{d\boldsymbol{X}(\phi)}{d\phi}[1 + \Omega(\phi)] = \boldsymbol{F}(\boldsymbol{X}(\phi)) + \varepsilon \boldsymbol{p}(\boldsymbol{X}(\phi)) . \tag{4.1.7}$$

Expanding $\boldsymbol{F}$ into a Taylor series,

$$\boldsymbol{F}(\boldsymbol{X}(\phi)) = \boldsymbol{F}(\boldsymbol{X}_0(\phi)) + L(\phi)\boldsymbol{u} + M(\phi)\boldsymbol{u}\boldsymbol{u} + \dots , \tag{4.1.8}$$

and substituting (4.1.4) into (4.1.7), we obtain perturbation equations for $\boldsymbol{u}$ and Ω in the form

$$\begin{aligned} & S(\phi)\,\Omega(\phi)\boldsymbol{u}_0 + \left[\frac{d}{d\phi} - L(\phi)\right]\boldsymbol{u} \\ & \quad = -\Omega(\phi)\frac{d\boldsymbol{u}}{d\phi} + M(\phi)\boldsymbol{u}\boldsymbol{u} + \varepsilon \boldsymbol{p}(\boldsymbol{X}_0 + \boldsymbol{u}) + O(|\boldsymbol{u}|^3) , \end{aligned} \tag{4.1.9}$$

where (3.4.5) has been used for the first term on the left-hand side.

The formulation becomes a little more transparent by working with $\tilde{\boldsymbol{u}}(\phi)$, defined by

$$\tilde{u}(\phi) = S^{-1}(\phi)u(\phi)\,, \tag{4.1.10}$$

rather than with u itself. Condition (4.1.5) then becomes

$$u_0^*\tilde{u}(\phi) = 0\,. \tag{4.1.11}$$

Applying $S^{-1}(\phi)$ to (4.1.9) from the left, and using the identity (3.4.3), we have

$$\Omega(\phi)u_0 + \left(\frac{d}{d\phi} - \Lambda\right)\tilde{u} = B(\phi)\,, \tag{4.1.12}$$

where

$$B(\phi) = -\Omega\cdot\left(S^{-1}\frac{dS}{d\phi} + \frac{d}{d\phi}\right)\tilde{u}$$
$$+ S^{-1}[MS\tilde{u}S\tilde{u} + \varepsilon p(X_0 + S\tilde{u})] + O(|\tilde{u}|^3)\,. \tag{4.1.13}$$

Let $\tilde{u}(\phi)$ be decomposed into eigenvectors:

$$\tilde{u}(\phi) = \sum_{l\neq 0} c_l(\phi)u_l\,,$$

where the restriction $l \neq 0$ comes from (4.1.11). We take the scalar product of each side of (4.1.12) with u_l^*. If $l = 0$, we have

$$\Omega(\phi) = u_0^* B(\phi)\,, \tag{4.1.14}$$

and if $l \neq 0$,

$$\left(\frac{d}{d\phi} - \lambda_l\right)c_l(\phi) = u_l^* B(\phi)\,. \tag{4.1.15}$$

The unknown quantities are $\Omega(\phi)$ and $c_l(\phi)$ $(l \neq 0)$, and we try to obtain them in the expansion form

$$\begin{pmatrix} \Omega(\phi) \\ \tilde{u}(\phi) \\ c_l(\phi) \\ B(\phi) \end{pmatrix} = \sum_{\nu=1}^{\infty} \varepsilon^{\nu} \begin{pmatrix} \Omega_\nu(\phi) \\ \tilde{u}_\nu(\phi) \\ c_{l\nu}(\phi) \\ B_\nu(\phi) \end{pmatrix}. \tag{4.1.16}$$

Then (4.1.14, 15) yield a system of balance equations

$$\Omega_\nu(\phi) = u_0^* B_\nu(\phi)\,, \tag{4.1.17}$$

$$\left(\frac{d}{d\phi} - \lambda_l\right)c_{l\nu}(\phi) = u_l^* B_\nu(\phi) \qquad (l \neq 0)\,. \tag{4.1.18}$$

These equations are integrated to give

$$c_{l\nu}(\phi) = e^{\lambda_l \phi} c_{l\nu}(0) + \int_0^\phi d\phi' e^{\lambda_l(\phi - \phi')} u_l^* B_\nu(\phi') \,. \tag{4.1.19}$$

It is easy to confirm from the general expression for B that $B_\nu(\phi)$ contains only the lower-order unknowns $\Omega_{\nu'}(\phi)$ and $c_{l\nu'}(\phi)$ $(\nu' < \nu)$, which says that (4.1.17, 18) represent a system of linear inhomogeneous equations.

Since we are only interested in the long-time behavior, we take the limit $\phi \to \infty$ (equivalent to $t \to \infty$). Then the first term on the right-hand side in (4.1.19) vanishes by virtue of the assumed property $\mathrm{Re}\{\lambda_l\} < 0$ $(l \neq 0)$. Furthermore, it can be proved that $\Omega_\nu(\phi)$ and $c_{l\nu}(\phi)$, for all ν, approach T-periodic functions of ϕ as $\phi \to \infty$. To show this, notice that $B_1(\phi)$ is T-periodic because

$$B_1(\phi) = S^{-1}(\phi)\Pi(\phi) \,. \tag{4.1.20}$$

Then, it follows from (4.1.17) that $\Omega_1(\phi)$ is T-periodic and, according to (4.1.19), $c_{l1}(\phi)$ also become T-periodic as $\phi \to \infty$. But if $\Omega_1(\phi)$ and $c_{l1}(\phi)$ are T-periodic, then this implies that $B_2(\phi)$ is T-periodic, which in turn implies that $\Omega_2(\phi)$ and $c_{l2}(\phi)$ are also T-periodic as $\phi \to \infty$. In the same manner, $\Omega_\nu(\phi)$ and $c_{l\nu}(\phi)$ may be shown to be T-periodic for all ν as $\phi \to \infty$. In most practical applications, the expressions for $\tilde{u}_\nu(\phi)$ are not important. What we actually need are $\Omega_\nu(\phi)$ for some lower values of ν, or an evolution equation in the form

$$\frac{d\phi}{dt} = 1 + \varepsilon\Omega_1(\phi) + \varepsilon^2\Omega_2(\phi) + \dots \,. \tag{4.1.21}$$

Note that the lowest-order result (3.4.10a) is recovered when (4.1.20) is substituted into (4.1.17) with $\nu = 1$.

In Sect. 3.2, $\Omega(\phi)$ was simply averaged through the formula (3.2.11). For higher-order terms, such a naive way of averaging the coefficients is no longer justified. A similar problem is encountered in the quasi-linear theory of nonlinear oscillations (Nayfeh, 1973). The correct averaging may be achieved by introducing a new phase variable $\bar{\phi}$ via

$$\phi = f(\bar{\phi}) = \bar{\phi} + \varepsilon a_1(\bar{\phi}) + \varepsilon^2 a_2(\bar{\phi}) + \dots \,, \tag{4.1.22}$$

where the $a_i(\bar{\phi})$ are T-periodic functions yet to be specified. This transformation is made unique by choosing some value, e.g., the zero value, to be a fixed point of the transformation $f(\bar{\phi})$ independently of ε. Thus, one is allowed to set

$$a_i(0) = 0 \,. \tag{4.1.23}$$

We require the new phase $\bar{\phi}$ to increase at a constant rate with t. Thus, $a_i(\bar{\phi})$ may be determined in such a way that $d\bar{\phi}/dt$ no longer depends on $\bar{\phi}$ itself. Let $d\bar{\phi}/dt$ be expressed as

$$\frac{d\bar{\phi}}{dt} = 1 + \varepsilon\omega_1 + \varepsilon^2\omega_2 + \dots , \tag{4.1.24}$$

where ω_i are constants. On the other hand, the substitution of (4.1.22) into (4.1.21) gives

$$\frac{d\bar{\phi}}{dt} = [1 + \varepsilon\Omega_1(\bar{\phi} + \varepsilon a_1(\bar{\phi}) + \dots) + \varepsilon^2\Omega_2(\bar{\phi} + \varepsilon a_1(\bar{\phi}) + \dots) + \dots] \times \left(1 + \varepsilon\frac{da_1}{d\bar{\phi}} + \varepsilon^2\frac{da_2}{d\bar{\phi}} + \dots\right)^{-1} . \tag{4.1.25}$$

Comparing (4.1.24) and (4.1.25), we obtain

$$\frac{da_\nu}{d\bar{\phi}} = b_\nu(\bar{\phi}) - \omega_\nu, \quad \text{where} \tag{4.1.26}$$

$$b_1(\bar{\phi}) = \Omega_1(\bar{\phi}) , \tag{4.1.27a}$$

$$b_2(\bar{\phi}) = \Omega_2(\bar{\phi}) + \frac{d\Omega_1}{d\bar{\phi}} a_1(\bar{\phi}) - \Omega_1(\bar{\phi})\frac{da_1}{d\bar{\phi}} + \left(\frac{da_1}{d\bar{\phi}}\right)^2 , \tag{4.1.27b}$$

etc. Note that the $b_\nu(\bar{\phi})$ depend only on the lower-order quantities $a_{\nu'}(\nu' < \nu)$. Since the solutions a_ν of (4.1.26) are T-periodic it is required that

$$\omega_\nu = \frac{1}{T}\int_0^T b_\nu(t)\,dt . \tag{4.1.28}$$

Equations (4.1.26) and (4.1.28) together with the initial conditions (4.1.23) are sufficient to determine all a_ν and ω_ν in an iterative way. In particular, we have

$$\omega_1 = \frac{1}{T}\int_0^T \Omega_1(t)\,dt , \tag{4.1.29}$$

which justifies the simple averaging to lowest order. In second order, such a naive averaging idea would lead to

$$\omega_2 = \frac{1}{T}\int_0^T \Omega_2(t)\,dt ,$$

which is incorrect, however. The calculation of ω_2 through (4.1.28) and (4.1.27b) actually leads to a much more complicated expression:

$$\omega_2 = \frac{1}{T}\int_0^T dt \left\{ \Omega_2(t) + \int_0^t dt' [\Omega_1(t') - \omega_1] \frac{d\Omega_1(t)}{dt} \right.$$
$$\left. - [\Omega_1(t) - \omega_1]\,\Omega_1(t) + [\Omega_1(t) - \omega_1]^2 \right\}. \tag{4.1.30}$$

4.2 Generalization of the Nonlinear Phase Diffusion Equation

We are now ready to take $\varepsilon \boldsymbol{p}(X)$ as representing a diffusion term, and to generalize the nonlinear phase diffusion equation (3.3.5). To say that $\nabla^2 \boldsymbol{X} \sim \varepsilon$ is equivalent to saying that the operator ∇ carries the smallness factor $\sqrt{\varepsilon}$, that is, whenever a spatial derivative appears, it generates a small quantity of order $\sqrt{\varepsilon}$. In fact, ∇ always appears as the combination $\sqrt{\varepsilon}\nabla$ in the theory below.

The formulation goes quite in parallel with that of Sect. 4.1. There is one important new feature, however: previously, we sought $\boldsymbol{u}$ and Ω as functions of $\phi(t)$ rather than of t itself. For the present system, these quantities depend on $\boldsymbol{r}$ as well as on t. A natural generalization of the previous functional idea suggests that $\boldsymbol{u}$ and Ω should be sought as functions of the entire spatial profile of ϕ and not simply of a single local value of ϕ. This suggest that $\boldsymbol{u}(\boldsymbol{r},t)$ and $\Omega(\boldsymbol{r},t)$ should be sought as functions of $\phi(\boldsymbol{r},t)$ and all its spatial derivatives. To indicate this kind of extended functional dependence, we use the notation $[\phi]$, or

$$A[\phi] \equiv A(\phi, \sqrt{\varepsilon}\nabla\phi, \varepsilon\nabla^2\phi, \ldots).$$

Our goal is to derive an evolution equation for ϕ in the form

$$\frac{\partial\phi}{\partial t} = 1 + \Omega[\phi], \tag{4.2.1}$$

which is similar to (4.1.6), but actually represents a *partial* differential equation.

We begin by splitting $\boldsymbol{X}[\phi]$, similarly to (4.1.4):

$$\boldsymbol{X}[\phi] = \boldsymbol{X}_0[\phi] + \boldsymbol{u}[\phi], \tag{4.2.2}$$

where $\boldsymbol{u}$ is supposed to satisfy

$$\boldsymbol{u}_0^* S^{-1}(\phi)\boldsymbol{u}[\phi] = 0, \tag{4.2.3}$$

analogously to (4.1.5). When we move from the old representation $A(\boldsymbol{r},t)$ to the new one $A[\phi]$ for some space-dependent quantity A, the time differentiation operating on A must be transformed as

$$\frac{\partial}{\partial t} \to \frac{\partial}{\partial\phi} + \sum_{j=0}^{\infty} (\sqrt{\varepsilon}\nabla)^j \Omega[\phi] \frac{\partial}{\partial(\sqrt{\varepsilon}\nabla)^j\phi}. \tag{4.2.4}$$

Thus our reaction-diffusion equations (with ε inserted before $D\nabla^2\boldsymbol{X}$) may be expressed as

$$\frac{\partial X[\phi]}{\partial \phi} + \sum_{j=0}^{\infty} (\sqrt{\varepsilon}\nabla)^j \Omega[\phi] \frac{\partial X[\phi]}{\partial (\sqrt{\varepsilon}\nabla)^j \phi} = F(X[\phi]) + \varepsilon D \nabla^2 X[\phi] . \tag{4.2.5}$$

Substituting (4.2.2) and the Taylor expansion of F in (4.1.8) into (4.2.5), we have

$$S(\phi)\,\Omega[\phi]u_0 + \left[\frac{\partial}{\partial \phi} - L(\phi)\right]u = -\sum_{j=0}^{\infty} (\sqrt{\varepsilon}\nabla)^j \Omega[\phi] \frac{\partial u}{\partial (\sqrt{\varepsilon}\nabla)^j \phi}$$
$$+ M(\phi)uu + \varepsilon D \nabla^2 (X_0 + u) + O(|u|^3) , \tag{4.2.6}$$

which should be compared to the corresponding equation (4.1.9).

Let $\tilde{u}[\phi]$ be defined by

$$\tilde{u}[\phi] = S^{-1}(\phi) u[\phi] . \tag{4.2.7}$$

Then $\tilde{u}$ must satisfy

$$u_0^* \tilde{u}[\phi] = 0 , \tag{4.2.8}$$

because of (4.2.3). We apply $S^{-1}(\phi)$ to (4.2.6) from the left to give an equation similar to (4.1.12):

$$\Omega[\phi]u_0 + \left(\frac{\partial}{\partial \phi} - \Lambda\right)\tilde{u} = B[\phi] , \quad \text{where} \tag{4.2.9}$$

$$B[\phi] = -\left[\Omega S^{-1}\frac{dS}{d\phi} + \sum_{j=0}^{\infty} (\sqrt{\varepsilon}\nabla)^j \Omega \frac{\partial}{\partial (\sqrt{\varepsilon}\nabla)^j \phi}\right]\tilde{u}$$
$$+ S^{-1}[M \cdot S\tilde{u}S\tilde{u} + \varepsilon D \nabla^2 (X_0 + S\tilde{u})] + O(|\tilde{u}|^3) . \tag{4.2.10}$$

Let $\tilde{u}[\phi]$ be decomposed as

$$\tilde{u}[\phi] = \sum_{l \neq 0} c_l[\phi] u_l . \tag{4.2.11}$$

We take a scalar product of each side of (4.2.9) with u_l^*. This gives

$$\Omega[\phi] = u_0^* B[\phi] \tag{4.2.12}$$

for $l = 0$, and

$$\left(\frac{\partial}{\partial \phi} - \lambda_l\right) c_l[\phi] = u_l^* B[\phi] \tag{4.2.13}$$

for $l \neq 0$. It is almost routine to expand various quantities in powers of ε as

$$\begin{pmatrix} \Omega[\phi] \\ \tilde{u}[\phi] \\ c_l[\phi] \\ B[\phi] \end{pmatrix} = \sum_{\nu=1}^{\infty} \varepsilon^{\nu} \begin{pmatrix} \Omega_\nu[\phi] \\ \tilde{u}_\nu[\phi] \\ c_{l\nu}[\phi] \\ B_\nu[\phi] \end{pmatrix}, \tag{4.2.14}$$

where $\tilde{u}_\nu$ and $c_{l\nu}$ are related to each other through

$$\tilde{u}_\nu[\phi] = \sum_{l \neq 0} c_{l\nu}[\phi] u_l. \tag{4.2.15}$$

One may alternatively make an expansion in powers of $\sqrt{\varepsilon}$ rather than ε. However, all odd powers of $\sqrt{\varepsilon}$ prove to vanish identically. In any case (4.2.12, 13) are reduced to

$$\Omega_\nu[\phi] = u_0^* B_\nu[\phi], \tag{4.2.16}$$

$$\left(\frac{\partial}{\partial \phi} - \lambda_l\right) c_{l\nu}[\phi] = u_l^* B_\nu[\phi], \quad l \neq 0. \tag{4.2.17}$$

Remember that ∇ carriers the smallness factor $\sqrt{\varepsilon}$. This means that the coefficient $A_\nu[\phi]$ of ε^ν in the ε-expansion of some quantity $A[\phi]$ must consist of terms in which the operator ∇ appears 2ν times in all possible combinations. For example,

$$\Omega_1[\phi] = \Omega_1^{(1)}(\phi) \nabla^2 \phi + \Omega_1^{(2)}(\phi)(\nabla \phi)^2, \tag{4.2.18}$$

$$\begin{aligned} \Omega_2[\phi] = {} & \Omega_2^{(1)}(\phi) \nabla^4 \phi + \Omega_2^{(2)}(\phi) \nabla^3 \phi \nabla \phi + \Omega_2^{(3)}(\phi)(\nabla^2 \phi)^2 \\ & + \Omega_2^{(4)}(\phi) \nabla^2 \phi (\nabla \phi)^2 + \Omega_2^{(5)}(\phi)(\nabla \phi)^4. \end{aligned} \tag{4.2.19}$$

As in these expressions, the coefficients of different types of terms for a given $A_\nu[\phi]$ will be specified by a superscript (σ): $A_\nu^{(\sigma)}(\phi)$. By sorting out various terms in (4.2.16, 17) in this manner, we obtain a finer set of balance equations:

$$\Omega_\nu^{(\sigma)}(\phi) = u_0^* B_\nu^{(\sigma)}(\phi), \tag{4.2.20}$$

$$c_{l\nu}^{(\sigma)}(\phi) = \int_{-\infty}^{\phi} d\phi' \, \mathrm{e}^{\lambda_l(\phi - \phi')} u_l^* B_\nu^{(\sigma)}(\phi). \tag{4.2.21}$$

These equations can be solved iteratively to give T-periodic $\Omega_\nu^{(\sigma)}(\phi)$ and $c_{l\nu}^{(\sigma)}(\phi)$, the reason for which is completely the same as in Sect. 4.1, and we do not repeat it.

Let us try to solve for some lower-order quantities. The inhomogeneous terms in lowest order are given by

$$B_1^{(1)}(\phi) = S^{-1}(\phi) D \frac{dX_0}{d\phi} = S^{-1}(\phi) D S(\phi) u_0 \tag{4.2.22a}$$

and

$$B_1^{(2)}(\phi) = S^{-1}(\phi)D\frac{d^2X_0}{d\phi^2} = S^{-1}(\phi)D\frac{dS(\phi)}{d\phi}u_0\,, \tag{4.2.22b}$$

which immediately lead to

$$\Omega_1^{(1)}(\phi) = u_0^* S^{-1}(\phi)DS(\phi)u_0\,, \tag{4.2.23a}$$

$$\Omega_1^{(2)}(\phi) = u_0^* S^{-1}(\phi)D\frac{dS(\phi)}{d\phi}u_0\,. \tag{4.2.23b}$$

Note that the above expressions coincide with the previous results (3.4.10b, c), respectively. The power of Method II becomes clear if we go to the next order. We will concentrate on obtaining $\Omega_2^{(1)}(\phi)$ since the corresponding $\nabla^4\phi$ terms turns out particularly important in the problem of turbulence. For simplicity, the following notations are introduced:

$$\tilde{D}(\phi) \equiv S^{-1}(\phi)DS(\phi)\,,$$

$$(l|A|m) \equiv u_l^* A u_m\,.$$

By inspection, we get from (4.2.10)

$$B_2^{(1)}(\phi) = \tilde{D}(\phi)\tilde{u}_1^{(1)}(\phi) = \sum_{l\neq 0}\tilde{D}(\phi)c_{l1}^{(1)}(\phi)u_l\,,$$

where the last equality follows from (4.2.15). The formula (4.2.20) now leads to

$$\Omega_2^{(1)}(\phi) = \sum_{l\neq 0}(0|\tilde{D}(\phi)|l)c_{l1}^{(1)}(\phi)\,. \tag{4.2.24}$$

We need the expression for $c_{l1}^{(1)}(\phi)$, which is provided by (4.2.21) with $B_1^{(1)}(\phi)$ given by (4.2.22a):

$$c_{l1}^{(1)}(\phi) = \int_{-\infty}^{\phi} d\phi' \mathrm{e}^{\lambda_l(\phi-\phi')}(l|\tilde{D}(\phi')|0)\,. \tag{4.2.25}$$

From (4.2.24, 25), we have

$$\Omega_2^{(1)}(\phi) = \sum_{l\neq 0}\int_{-\infty}^{\phi} d\phi'(0|\tilde{D}(\phi)|l)(l|\tilde{D}(\phi')|0)\mathrm{e}^{\lambda_l(\phi-\phi')}\,. \tag{4.2.26}$$

In Sect. 3.5, we calculated explicitly $\Omega_1^{(1)}$ and $\Omega_1^{(2)}$ for the Ginzburg-Landau equation. It is now possible to calculate $\Omega_2^{(1)}$ for the same model. Noting that $\tilde{D}$ is independent of ϕ for this particular model, and using the eigenvectors and eigenvalues obtained in Sect. 3.5, we easily get

$$\Omega_2^{(1)} = -c_1^2(1+c_2^2)/2 . \tag{4.2.27}$$

Since this is non-positive, the $\nabla^4\phi$ term represents a damping, while the phase diffusion coefficient $\Omega_1^{(1)}$ may be negative, causing instability.

According to the present method, the evolution equation for ϕ generally assumes the form

$$\frac{\partial\phi}{\partial t} = 1+\varepsilon[\Omega_1^{(1)}(\phi)\,\nabla^2\phi+\Omega_1^{(2)}(\phi)(\nabla\phi)^2]+\varepsilon^2[\Omega_2^{(1)}(\phi)\,\nabla^4\phi+\dots$$
$$+\Omega_2^{(5)}(\phi)(\nabla\phi)^4]+\dots , \tag{4.2.28}$$

which should be contrasted with (4.1.21). How to average correctly is now described. As an extension of (4.1.22), it is appropriate to introduce $\bar{\phi}$ through

$$\phi = f[\bar{\phi}] = \bar{\phi}+\varepsilon[a_1^{(1)}(\bar{\phi})\,\nabla^2\bar{\phi}+a_1^{(2)}(\bar{\phi})(\nabla\bar{\phi})^2]+\varepsilon^2[a_2^{(1)}(\bar{\phi})\,\nabla^4\bar{\phi}+\dots$$
$$+a_2^{(5)}(\bar{\phi})(\nabla\bar{\phi})^4]+\dots . \tag{4.2.29}$$

The new variable $\bar{\phi}$ is required to obey an equation of constant coefficients, i.e.

$$\frac{\partial\bar{\phi}}{\partial t} = 1+\varepsilon[\omega_1^{(1)}\,\nabla^2\bar{\phi}+\omega_1^{(2)}(\nabla\bar{\phi})^2]+\varepsilon^2[\omega_2^{(1)}\,\nabla^4\bar{\phi}+\dots+\omega_2^{(5)}(\nabla\bar{\phi})^4]+\dots . \tag{4.2.30}$$

On the other hand, the substitution of (4.2.29) into (4.2.28) gives the equation for $d\bar{\phi}/dt$ which, however, is too cumbersome to write down. Its comparison with (4.2.30) for terms in ε yields a system of equations in the form

$$\frac{da_\nu^{(\sigma)}}{d\bar{\phi}} = b_\nu^{(\sigma)}(\bar{\phi})-\omega_\nu^{(\sigma)} . \tag{4.2.31}$$

The first few $b_\nu^{(\sigma)}(\bar{\phi})$ are rather easy to write down:

$$b_1^{(1)}(\bar{\phi}) = \Omega_1^{(1)}(\bar{\phi}) , \tag{4.2.32a}$$

$$b_1^{(2)}(\bar{\phi}) = \Omega_1^{(2)}(\bar{\phi}) , \tag{4.2.32b}$$

$$b_2^{(1)}(\bar{\phi}) = \Omega_2^{(1)}(\bar{\phi})+a_1^{(1)}(\bar{\phi})[\Omega_1^{(1)}(\bar{\phi})-\omega_1^{(1)}] . \tag{4.2.32c}$$

Completely the same reasoning as in Sect. 4.1 can be applied, leading to

$$\omega_\nu^{(\sigma)} = \frac{1}{T}\int_0^T b_\nu^{(\sigma)}(t)\,dt . \tag{4.2.33}$$

In particular,

$$\omega_1^{(\sigma)} = \frac{1}{T}\int_0^T \Omega_1^{(\sigma)}(t)\,dt , \qquad \sigma = 1,2 , \tag{4.2.34}$$

which coincides with α and β in (3.3.7). The next-order quantity $\omega_2^{(1)}$ is found to be

$$\omega_2^{(1)} = \frac{1}{T}\int_0^T dt \left\{ \Omega_2^{(1)}(t) + [\Omega_1^{(1)}(t) - \omega_1^{(1)}] \int_0^t [\Omega_1^{(1)}(t') - \omega_1^{(1)}] dt' \right\}. \tag{4.2.35}$$

For later convenience, we use informal notations α, β, and $-\gamma$ for $\omega_1^{(1)}$, $\omega_1^{(2)}$, and $\omega_2^{(1)}$, respectively. After setting ε equal to 1, we write (4.2.30) as

$$\frac{\partial \psi}{\partial t} = \alpha \nabla^2 \psi + \beta (\nabla \psi)^2 - \gamma \nabla^4 \psi + \dots , \tag{4.2.36}$$

where $\psi \equiv \bar{\phi} - t$. We shall often refer to the form (4.2.36) in later chapters.

The dispersion relation

$$\lambda = -\alpha k^2 - \gamma k^4 + \dots \tag{4.2.37}$$

is implied from the linearization of (4.2.36) and by making the correspondence $(\partial/\partial t, \nabla) \leftrightarrow (\lambda, ik)$. Equation (4.2.37) is, in fact, considered to give a correct expansion form of the linear dispersion associated with the phaselike fluctuations about the homogeneous oscillation $X_0(t)$ (for more details, see Sect. 7.2).

4.3 Dynamics of Slowly Varying Wavefronts

Having thus reformulated Method I into a more systematic form, we now notice that the variety of problems that the theory can treat has become considerably richer. This and the next sections take up two such examples that might possibly have been overlooked without Method II. Both are related to typical phenomena in reaction-diffusion systems but are not necessarily of an oscillatory nature.

The first class of phenomena we would like to discuss concerns the dynamics of wavefronts of certain types of chemical waves in two dimensional media. (Extension to three-dimensional cases is straightforward, and not discussed here.) Let x and y denote the spatial coordinates. Imagine at first that the composition vector X has no y-dependence, so that we are essentially working with a one-dimensional system:

$$\frac{\partial X}{\partial t} = F(X) + D\frac{\partial^2 X}{\partial x^2}. \tag{4.3.1}$$

The system length is supposed to be infinite. Without diffusion, the system is assumed to admit one or more stable equilibrium points but no stable oscillations. Furthermore, these equilibrium states are supposed to remain stable to nonuniform fluctuations. As is well known, it sometimes happens that steadily traveling nonlinear waves arise in such non-oscillatory media. (For a special model, see Sect. 7.3.) Two types of waves are possible, which are schematically

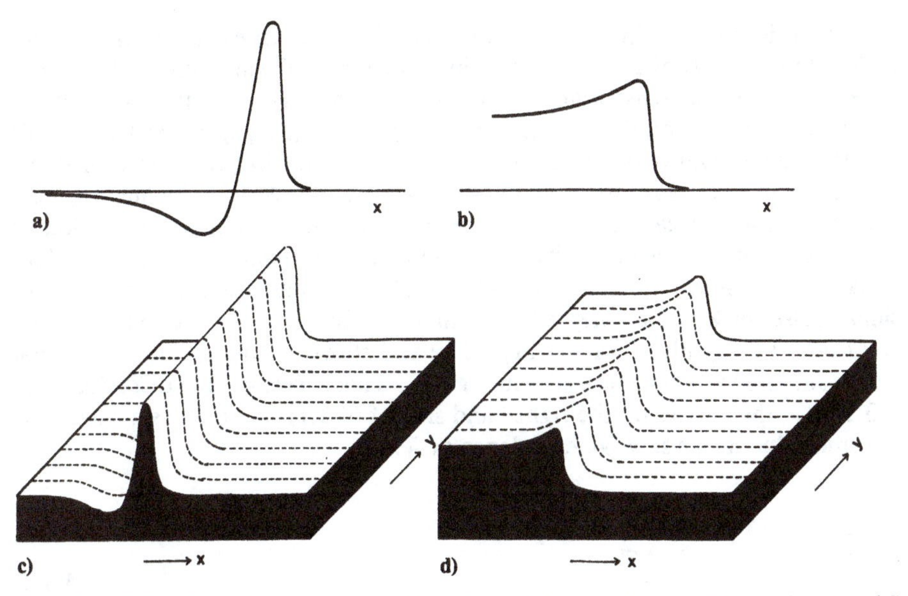

Fig. 4.2a – d. One-dimensional pulse (**a**) and kink (**b**), and their two-dimensional extensions (**c** and **d**)

shown in Fig. 4.2. We call the first type (Fig. 4.2a) a pulse, and the second one (Fig. 4.2b) a kink. Although the latter is often called a front, we do not use this term because we want to reserve a similar term "wavefront" for describing a certain suitably defined locus for both types of waves. Pulses are also called trigger waves, and they arise in media possessing excitability, such as a non-oscillatory version of the Belousov-Zhabotinsky reaction system and nerve axons. For mathematical work on trigger waves in the Belousov-Zhabotinsky reaction (not necessarily the oscillationless version), see Murray (1976), Karfunkel and Seelig (1975), Karfunkel and Kahlert (1977), Hastings (1976), and the lecture notes of Tyson (1976). Pulses in some nerve conduction equations were studied by FitzHugh (1961, 1969), Nagumo et al. (1962), McKean (1970), Rinzel and Keller (1973), Rinzel (1975), and many other people. Pulses often travel as a wave train, but we will only consider a single isolated wave in this section; a pulse train will be considered in Sect. 4.4., but in a slightly different context. The equilibrium states extending before and behind a pulse correspond to an identical equilibrium solution of the diffusionless equations. In contrast, kinks are generally possible when the corresponding diffusionless systems have multiple equilibrium states; a kink represents a narrow transition region from one equilibrium state to the other. Although there exist a number of reaction-diffusion models which may show kinks, one has never been able to visually produce them in chemical reaction experiments. A feature common to pulses and kinks is that the medium is quiescent almost everywhere while a sharp spatial variation in X is steadily maintained in a very narrow region.

Method II becomes relevant to the dynamics of such chemical waves when they are extended to form two-dimensional waves. Let $X_0(x-ct)$ denote a steadily traveling pulse- or kink solution of (4.3.1), where c is the propagation velocity. It is worth noting that this solution is in some sense analogous to a limit cycle solution $X_0(t)$ of a system of ordinary differential equations. In fact, both have "phase" in the sense that the time translation $t \to t + \psi_0$, with arbitrary ψ_0, again yields steady solutions $X_0(x - c(t + \psi_0))$ and $X_0(t + \psi_0)$, respectively. This says that there always exists a zero eigenvalue (corresponding to translational disturbances) in the linearized system about $X_0(x - ct)$ or $X_0(t)$. The analogy may persist when some weak coupling is introduced between such modes of motion. It may be evident what is meant by interaction between oscillators. But what does weak coupling mean for pulses and kinks? This actually means that one-dimensional pulses or kinks are aligned to form a two-dimensional wave whose wave-front deviates slightly from a straight line and shows a slow undulatory spatial variation, as shown schematically in Fig. 4.3. This is because the diffusion term $D\partial^2 X/\partial y^2$ then arises on the right-hand side of (4.3.1), but only as a small perturbation due to the slow y-dependence of X:

$$\frac{\partial X}{\partial t} = F(X) + D\frac{\partial^2 X}{\partial x^2} + \varepsilon p\,, \qquad p = D\frac{\partial^2 X}{\partial y^2}\,. \tag{4.3.2}$$

It would be inappropriate if not absolutely impossible, to apply Method I to problem (4.3.2), whereas Method II seems to be much more suited to it. In fact, we shall find a perfect analogy existing between the formulation below and that of Sect. 4.2.

We introduce some notions associated with the linearization of the unperturbed system (i.e., the one-dimensional system) about the traveling solution $X_0(x - c(t + \psi_0))$. Let z denote a moving coordinate defined by $z = x - c(t + \psi_0)$. The traveling solution $X_0(z)$ has to satisfy

$$F(X_0) + c\frac{dX_0}{dz} + D\frac{d^2 X_0}{dz^2} = 0\,. \tag{4.3.3}$$

Let $u(z, t)$ represent disturbances defined by

$$X(z, t) = X_0(z) + u(z, t)\,.$$

Equation (4.3.1) is linearized in u to give

$$\frac{\partial u}{\partial t} = \Gamma(z)u\,, \quad \text{where} \tag{4.3.4}$$

$$\Gamma(z) = L(z) + c\frac{\partial}{\partial z} + D\frac{\partial^2}{\partial z^2}\,. \tag{4.3.5}$$

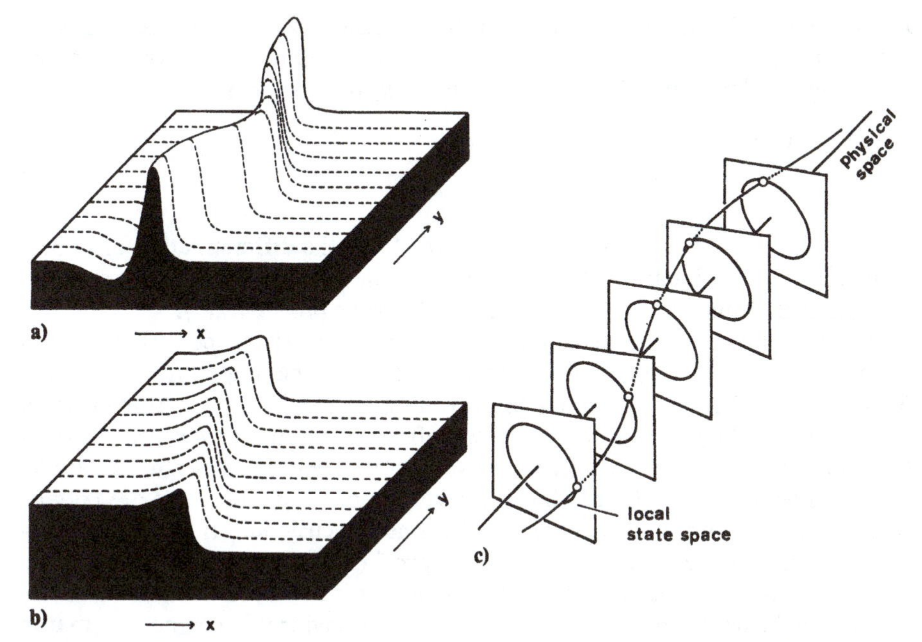

Fig. 4.3 a – c. Deformed wavefronts of pulse type (**a**) and kink type (**b**), and their analogue to a chain of coupled oscillators with spatial phase variation (**c**)

Let u_l and λ_l denote the eigenvectors and eigenvalues (assumed to be simple) of $\Gamma(z)$, i.e., $\Gamma(z)u_l = \lambda_l u_l$, $l = 0, 1, 2, \ldots$. Similarly, the eigenvectors of the adjoint operator $\Gamma^*(z)$ of $\Gamma(z)$ belonging to λ_l are denoted as u_l^*, i.e., $\Gamma^*(z)u_l^* = \lambda_l u_l^*$. Here $\Gamma^*(z)$ is defined by

$$\int_{-\infty}^{\infty} f(z)\Gamma g(z)\,dz = \int_{-\infty}^{\infty} g(z)\Gamma^* f(z)\,dz\,,$$

where f and g may be arbitrary vector functions of z which make the above integrals sensible. Specifically, $\Gamma^*(z)$ is expressed as

$$\Gamma^* = {}^{\mathrm{t}}L(z) - c\frac{d}{dz} + {}^{\mathrm{t}}D\frac{d^2}{dz^2}\,, \tag{4.3.6}$$

where the superscript t indicates the transpose. The eigenvectors are supposed to be orthonormalized as

$$\int_{-\infty}^{\infty} u_l^*(z)u_m(z)\,dz = \delta_{lm}\,. \tag{4.3.7}$$

As noted above, there exists at least one eigenvalue, say λ_0, which vanishes identically. We assume all the other eigenvalues lie in the left half of the complex λ plane and at a finite distance from the imaginary axis. The assumption that the zero eigenvalue is isolated is justified in the present single pulse (kink) problem, but by no means valid for periodic wave trains. Quite analogously to the foregoing oscillator case, the zero eigenvector $\boldsymbol{u}_0$ may be chosen as

$$\boldsymbol{u}_0 = \frac{dX_0}{dz}. \tag{4.3.8}$$

In fact, the identity $\Gamma(z)dX_0/dz = 0$ follows from the differentiation of (4.3.3) with respect to z. Equation (4.3.8) shows that the zero eigenvector is associated with an infinitesimal spatial translation of the wave profile $X_0(z)$.

We now move to the perturbed system. The underlying perturbation idea is the following: If the space dimension is two, the unperturbed solution $X_0(x - c(t + \psi_0))$ represents a steadily traveling wave whose wavefront forms a straight line in the y direction. Let the wavefront be given a slow wavy deformation as in Fig. 4.3a or b. Locally the wavefront is supposed to be almost parallel to the y direction everywhere, while the deformation amplitude itself need not be small. The situation here reminds us of a one-dimensional array of diffusion-coupled limit cycle oscillators whose phase profile shows a slow spatial variation (Fig. 4.3c). We have seen that each local oscillator is not much disturbed from its unperturbed closed orbit as far as the long-time behavior is concerned, so that in the crudest approximation one could assume

$$X(x,t) \simeq X_0(\phi(x,t)).$$

This idea may readily be carried over to the wavefront problem. In fact, the local wave profile seen along the direction vertical to the wavefront may be assumed to be almost the same as that of the unperturbed system, while the phase, i.e., the locus of the wavefront, is allowed to vary slowly with y. This is equivalent to saying that

$$X(x,y,t) \simeq X_0(\zeta), \quad \text{where} \tag{4.3.9}$$

$$\zeta = x - c\phi(y,t),$$

with $\phi(y,t)$ slowly varying in y (and possibly also in t). In a more precise picture, the y-dependence of ϕ generally causes some deformation of the wave profile itself, just as an oscillator, if coupled to its neighbors, can no longer stay exactly on its natural orbit. By taking this effect into account, (4.3.9) is generalized to the following system of equations:

$$X(x,y,t) = X(\zeta,[\phi]) = X_0(\zeta) + \boldsymbol{u}(\zeta,[\phi]), \tag{4.3.10}$$

$$\frac{\partial\phi}{\partial t} = 1 + \Omega[\phi]. \tag{4.3.11}$$

The former equation is analogous to (4.2.2), and the latter is identical to (4.2.1). In the above, notations such as $A(\zeta,[\phi])$ and $A[\phi]$ should be understood as

$$A(\zeta,[\phi]) \equiv A\left(\zeta, \sqrt{\varepsilon}\frac{\partial\phi}{\partial y}, \varepsilon\frac{\partial^2\phi}{\partial y^2}, \ldots\right),$$
$$A[\phi] = A\left(\sqrt{\varepsilon}\frac{\partial\phi}{\partial y}, \varepsilon\frac{\partial^2\phi}{\partial y^2}, \ldots\right); \tag{4.3.12}$$

note that ϕ itself does not appear except through ζ, a property which is expected from the translational symmetry of the system. Actually, the omission of explicit ϕ-dependence also has the effect of washing out the initial transients automatically.

Let (4.3.10) be substituted into (4.3.2). This yields

$$-c\Omega[\phi]\boldsymbol{u}_0 - \Gamma(\zeta)\boldsymbol{u}(\zeta,[\phi]) = \boldsymbol{B}(\zeta,[\phi]), \quad \text{where} \tag{4.3.13}$$

$$\boldsymbol{B}(\zeta,[\phi]) = -\sum_{j=0}^{\infty}(\sqrt{\varepsilon}\partial_y)^j\Omega\frac{\partial\boldsymbol{u}}{\partial(\sqrt{\varepsilon}\partial_y)^j\phi} + \boldsymbol{M}(\zeta)\boldsymbol{u}\boldsymbol{u} + \varepsilon D\partial_y^2(X_0+\boldsymbol{u}) + O(|\boldsymbol{u}|^3), \tag{4.3.14}$$

with $\partial_y \equiv \partial/\partial y$. Let $\boldsymbol{u}$ be decomposed into eigenvectors as

$$\boldsymbol{u}(\zeta,[\phi]) = \sum_{l\neq 0} c_l[\phi]\boldsymbol{u}_l(\zeta), \tag{4.3.15}$$

and substituted into (4.3.13). On taking a scalar product of (4.3.13) with $\boldsymbol{u}_l^*(\zeta)$ and integrating with respect to ζ over $(-\infty,\infty)$, we get

$$-c\Omega[\phi] = \int_{-\infty}^{\infty}\boldsymbol{u}_0^*(\zeta)\boldsymbol{B}(\zeta,[\phi])d\zeta \tag{4.3.16}$$

for $l=0$, and

$$-\lambda_l c_l[\phi] = \int_{-\infty}^{\infty}\boldsymbol{u}_l^*(\zeta)\boldsymbol{B}(\zeta,[\phi])d\zeta \tag{4.3.17}$$

for $l\neq 0$. Various quantities are now sought in the form of an ε expansion:

$$\begin{pmatrix}\Omega[\phi]\\ \boldsymbol{u}(\zeta,[\phi])\\ c_l[\phi]\\ \boldsymbol{B}(\zeta,[\phi])\end{pmatrix} = \sum_{\nu=1}^{\infty}\varepsilon^\nu\begin{pmatrix}\Omega_\nu[\phi]\\ \boldsymbol{u}_\nu(\zeta,[\phi])\\ c_{l\nu}[\phi]\\ \boldsymbol{B}_\nu(\zeta,[\phi])\end{pmatrix}. \tag{4.3.18}$$

Then (4.3.16, 17) become equivalent to

$$-c\Omega_\nu[\phi] = \int_{-\infty}^{\infty}\boldsymbol{u}_0^*(\zeta)\boldsymbol{B}_\nu(\zeta,[\phi])d\zeta, \tag{4.3.19}$$

$$-\lambda_l c_{l\nu}[\phi] = \int_{-\infty}^{\infty} u_l^*(\zeta) B_\nu(\zeta, [\phi]) d\zeta . \tag{4.3.20}$$

Each term on the right-hand side of (4.3.18) is further decomposed:

$$\Omega_1[\phi] = \Omega_1^{(1)} \frac{\partial^2 \phi}{\partial y^2} + \Omega_1^{(2)} \left(\frac{\partial \phi}{\partial y}\right)^2 , \tag{4.3.21}$$

$$\Omega_2[\phi] = \Omega_2^{(1)} \frac{\partial^4 \phi}{\partial y^4} + \ldots + \Omega_2^{(5)} \left(\frac{\partial \phi}{\partial y}\right)^4 , \quad \text{etc.} \tag{4.3.22}$$

Note that all coefficients on the right-hand sides of these equations are independent of ϕ, in contrast to the previous oscillator case, and this saves us cumbersome averaging procedures like those in Sect. 4.2. Equations (4.3.19, 20) may now be decomposed into finer balance equations:

$$-c\Omega_\nu^{(\sigma)} = \int_{-\infty}^{\infty} u_0^*(\zeta) B_\nu^{(\sigma)}(\zeta) d\zeta , \tag{4.3.23}$$

$$-\lambda_l c_{l\nu}^{(\sigma)} = \int_{-\infty}^{\infty} u_l^*(\zeta) B_\nu^{(\sigma)}(\zeta) d\zeta . \tag{4.3.24}$$

For the lowest-order inhomogeneous terms, we have

$$B_1^{(1)}(\zeta) = -cD \frac{dX_0}{d\zeta} = -cDu_0 , \tag{4.3.25 a}$$

$$B_1^{(2)}(\zeta) = c^2 D \frac{d^2 X_0}{d\zeta^2} = c^2 D \frac{du_0}{d\zeta} . \tag{4.3.25 b}$$

Thus,

$$\Omega_1^{(1)} = (0|D|0) \equiv \alpha , \tag{4.3.26 a}$$

$$\Omega_1^{(2)} = -c\left(0\left|D\frac{d}{d\zeta}\right|0\right) \equiv \beta , \tag{4.3.26 b}$$

where we have used notations like

$$(l|A|m) \equiv \int_{-\infty}^{\infty} u_l^*(\zeta) A u_m(\zeta) d\zeta .$$

The important second-order quantity $\Omega_2^{(1)}$ is also easy to compute. It has a form familar to us, appearing in many second-order perturbation theories:

$$\Omega_2^{(1)} = -\sum_{l \neq 0} \lambda_l^{-1} (0|D|l)(l|D|0) \equiv -\gamma . \tag{4.3.27}$$

Clearly, the wavefront equation takes the form

$$\frac{\partial \psi}{\partial t} = \alpha \frac{\partial^2 \psi}{\partial y^2} + \beta \left(\frac{\partial \psi}{\partial y} \right)^2 - \gamma \frac{\partial^4 \psi}{\partial y^4} + \dots , \tag{4.3.28}$$

where $\psi = \phi - t$. The above is identical to (4.2.36) in one space dimension.

Finally, we note the identity

$$\beta = c/2 \tag{4.3.29}$$

which is proved as follows: Let the equation $\Gamma(z)\boldsymbol{u}_0 = 0$ be written explicitly as

$$\left[L(z) + c\frac{d}{dz} + D\frac{d^2}{dz^2} \right] \boldsymbol{u}_0(z) = 0 . \tag{4.3.30}$$

Similarly, we have for $\Gamma^*(z)\boldsymbol{u}_0^* = 0$

$$\left[{}^{\mathrm{t}}L(z) - c\frac{d}{dz} + {}^{\mathrm{t}}D\frac{d^2}{dz^2} \right] \boldsymbol{u}_0^*(z) = 0 ,$$

or by taking its transpose,

$$\boldsymbol{u}_0^*(z) L(z) - c\frac{d}{dz}\boldsymbol{u}_0^*(z) + \frac{d^2}{dz^2}\boldsymbol{u}_0^* D = 0 . \tag{4.3.31}$$

We subtract the product formed by left multiplication of (4.3.30) by $\boldsymbol{u}_0^*$ from the product formed by right multiplication of (4.3.31) by $\boldsymbol{u}_0$, and obtain

$$\left(\frac{d^2}{dz^2}\boldsymbol{u}_0^* D \boldsymbol{u}_0 - \boldsymbol{u}_0^* D \frac{d^2}{dz^2}\boldsymbol{u}_0 \right) - c\left(\frac{d}{dz}\boldsymbol{u}_0^* \boldsymbol{u}_0 + \boldsymbol{u}_0^* \frac{d}{dz}\boldsymbol{u}_0 \right) = 0 , \tag{4.3.32}$$

or equivalently,

$$\frac{d}{dz}\left[\left(\frac{d}{dz}\boldsymbol{u}_0^* D \boldsymbol{u}_0 - \boldsymbol{u}_0^* D \frac{d}{dz}\boldsymbol{u}_0 \right) - c\boldsymbol{u}_0^* \boldsymbol{u}_0 \right] = 0 . \tag{4.3.33}$$

One may reasonably assume that $\boldsymbol{u}_0(z)$ and $\boldsymbol{u}_0^*(z)$ go to zero sufficiently fast as z goes to $\pm\infty$, so that (4.3.33) may be integrated to give

$$\frac{d}{dz}\boldsymbol{u}_0^* D \boldsymbol{u}_0 - \boldsymbol{u}_0^* D \frac{d\boldsymbol{u}_0}{dz} = c\boldsymbol{u}_0^* \boldsymbol{u}_0 . \tag{4.3.34}$$

Integrating once again, we have

$$c = \int_{-\infty}^{\infty} \left(\frac{d}{dz}\boldsymbol{u}_0^* D \boldsymbol{u}_0 - \boldsymbol{u}_0^* D \frac{d}{dz}\boldsymbol{u}_0 \right) dz$$

$$= -2 \int_{-\infty}^{\infty} u_0^* D \frac{d}{dz} u_0 dz \,. \tag{4.3.35}$$

The last expression is identical to 2β, according to (4.3.26b), which is what we wanted to prove.

4.4 Dynamics of Slowly Phase-Modulated Periodic Waves

In excitable reaction-diffusion systems, pulses can travel as a periodic wave train. In oscillatory reaction-diffusion systems, too, the existence of plane wave solutions has been theoretically established (Kopell and Howard, 1973a). In this section we will be concerned with such periodic waves in one space dimension, particularly when the local wavenumber slowly and slightly varies with x. For these systems, the analogy to systems of weakly coupled oscillators might look even weaker. Actually, however, there exists a rather strong formalistic similarity between the two.

When pulses (or peaks of plane waves) are distributed to form an aperiodic train, they are expected to regulate themselves to reestablish a perfect periodicity. Under certain conditions, such processes may be described by means of Method II. In the lowest non-trivial approximation, the evolution equation for the number density $n(x,t)$ of the pulses turns out identical to the Burgers equation in a suitable moving coordinate. In this connection, there exists an interesting experimental study concerning the propagation of randomly generated pulses of action potential along a giant axon of a squid (Musha et al., 1981). Although we do not attempt here a theoretical explanation of the remarkable statistical properties observed there for $n(x,t)$, the Burgers equation for n, if analyzed as a random boundary value problem, might possibly shed light on some mechanisms underlying peculiar spectral characteristics observed.

The argument below may apply equally to the pulses in excitable media and to smoother waves in oscillatory media. For simplicity, however, we shall use terms appropriate for pulse systems. Suppose the interpulse distances l are not uniformly distributed, but fluctuate slightly about some mean value $\bar{l}$ (Fig. 4.4). If we regard l as a function of x, the short-wavelength fluctuations of l will be quickly smoothed out. One may then expect the existence of a time stage where only the fluctuations whose typical wavelength is much longer than $\bar{l}$ survive. We want to learn how such fluctuations evolve in time. Since there exist two characteristic lengths, as suggested above, we introduce a new scale ξ through

$$\xi = \varepsilon x \,,$$

and reinterpret the solution $X(x,t)$ of (4.3.1) as depending on ξ and t, as well as its $\bar{l}$-periodic dependence on x:

$$X(x,t) \to X(x,\xi,t) = X(x+\bar{l},\xi,t) \,. \tag{4.4.1}$$

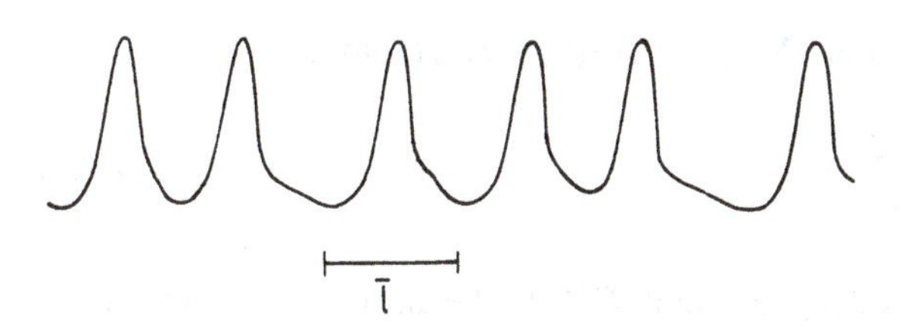

Fig. 4.4. Pulse train with non-uniform interpulse spacing

Here again, x and ξ are treated mathematically as independent variables. Correspondingly, the spatial differentiation appearing in (4.3.1) has to be transformed as

$$\frac{\partial}{\partial x} \rightarrow \frac{\partial}{\partial x} + \varepsilon \frac{\partial}{\partial \xi} . \tag{4.4.2}$$

Equation (4.3.1) now takes the form

$$\frac{\partial X}{\partial t} = F(X) + D\left(\frac{\partial}{\partial x} + \varepsilon \frac{\partial}{\partial \xi}\right)^2 X . \tag{4.4.3}$$

The terms of $O(\varepsilon)$ and $O(\varepsilon^2)$ have to be treated as a perturbation. By the term "unperturbed system", we therefore mean a strictly periodic system, i.e., system (4.3.1) subject to the periodic condition

$$X(x+\bar{l}, t) = X(x, t) . \tag{4.4.4}$$

Let z denote a moving coordinate, defined by

$$z = x - c(t + \psi_0) ,$$

and $X_0(z)$ a steadily traveling periodic solution of (4.3.1) under the condition (4.4.4):

$$F(X_0) + c\frac{dX_0}{dz} + D\frac{d^2X_0}{dz^2} = 0 , \tag{4.4.5}$$

$$X_0(z+\bar{l}) = X_0(z) . \tag{4.4.6}$$

Usually, we have a family of such periodic waves with various $\bar{l}$ (Rinzel and Keller, 1973). Here we consider the problem for a given value of $\bar{l}$. The linearization of the unperturbed system about X_0 leads to

$$\frac{du}{dt} = \Gamma(z)u , \tag{4.4.7a}$$

$$u(z+\bar{l}, t) = u(z, t) , \tag{4.4.7b}$$

where $u(z,t)$ is the deviation, i.e., $X(x,t)=X_0(z)+u(z,t)$, and

$$\Gamma(z)=L(z)+c\frac{\partial}{\partial z}+D\frac{\partial^2}{\partial z^2}\,. \tag{4.4.8}$$

We use the notations $u_l(z)$, $u_l^*(z)$, λ_l, and $\Gamma^*(z)$, the meanings of which are similar to those in Sect. 4.3. Note, however, that u_l, u_l^*, Γ, and Γ^* are all $\bar{l}$-periodic functions of z in the present case, so that the orthonormality conditions should be given by

$$\int_0^{\bar{l}} u_l^*(z)u_m(z)\,dz=\delta_{lm}\,.$$

If we abandoned the periodic condition, and considered the eigenvalue problem for systems of infinite length, then the eigenvalue spectrum would no longer be discrete; then no perturbation theories like Method II would be applicable. The reason for the existence of a zero eigenvalue is obvious. Again, the corresponding eigenvector is associated with an infinitesimal translation of $X_0(z)$, i.e.,

$$u_0(z)=\frac{dX_0}{dz}\,. \tag{4.4.9}$$

Having thus introduced some notions about the unperturbed system, we come back to the system (4.4.3) with perturbation. In analogy to the previous section, the solution of slowly phase-modulated waves may be sought in the form

$$X(x,t)=X_0(\zeta)+u(\zeta,[\phi])\,, \tag{4.4.10}$$

$$\frac{\partial\phi}{\partial t}=1+\Omega[\phi]\,,\qquad\text{where} \tag{4.4.11}$$

$$\zeta=x-c\phi(\zeta,t)$$

and u is periodic, i.e.,

$$u(\zeta+\bar{l},[\phi])=u(\zeta,[\phi])\,, \tag{4.4.12}$$

which satisfies the orthogonality condition with the zero eigenvector:

$$\int_0^{\bar{l}} u_0^*(\zeta)u(\zeta,[\phi])\,d\zeta=0\,. \tag{4.4.13}$$

The meaning of the notation $[\phi]$ is the same as in (4.3.12) with y replaced by ξ. From (4.4.3, 10),

$$-c\Omega[\phi]u_0-\Gamma(\zeta)u(\zeta,[\phi])=B(\zeta,[\phi])\,, \tag{4.4.14}$$

where

$$B(\zeta,[\phi]) = -\sum_{j=0}^{\infty}(\sqrt{\varepsilon}\partial_\xi)^j \Omega[\phi]\frac{\partial u}{\partial(\sqrt{\varepsilon}\partial_\xi)^j\phi} + M(\zeta)uu$$
$$+\sqrt{\varepsilon}D\frac{\partial}{\partial\xi}\left(2\frac{\partial}{\partial x}+\sqrt{\varepsilon}\frac{\partial}{\partial\xi}\right)(X_0+u)+O(|u|^3)\,. \qquad (4.4.15)$$

Let u be expressed in terms of the eigenvectors, similarly to (4.3.15). Multiplying (4.4.14) by $u_l^*(\zeta)$ and integrating over ζ, we obtain

$$-c\Omega[\phi] = \int_0^T u_0^*(\zeta)B(\zeta,[\phi])d\zeta\,, \qquad (4.4.16)$$

$$-\lambda_l c_l[\phi] = \int_0^T u_l^*(\zeta)B(\zeta,[\phi])d\zeta\,, \quad l \neq 0\,. \qquad (4.4.17)$$

Various quantities are expanded in powers of ε:

$$\begin{pmatrix}\Omega[\phi]\\ u(\zeta,[\phi])\\ c_l[\phi]\\ B(\zeta,[\phi])\end{pmatrix} = \sum_{\nu=1}^{\infty}\varepsilon^\nu\begin{pmatrix}\Omega_\nu[\phi]\\ u_\nu(\zeta,[\phi])\\ c_{l\nu}[\phi]\\ B_\nu(\zeta,[\phi])\end{pmatrix}.$$

The general fine structure of each expansion term on the right-hand side may be understood from the following few examples:

$$\Omega_1[\phi] = \Omega_1^{(1)}\frac{\partial\phi}{\partial\xi}\,,$$

$$\Omega_2[\phi] = \Omega_2^{(1)}\frac{\partial^2\phi}{\partial\xi^2}+\Omega_2^{(2)}\left(\frac{\partial\phi}{\partial\xi}\right)^2,$$

$$\Omega_3[\phi] = \Omega_3^{(1)}\frac{\partial^3\phi}{\partial\xi^3}+\Omega_3^{(2)}\frac{\partial^2\phi}{\partial\xi^2}\frac{\partial\phi}{\partial\xi}+\Omega_3^{(3)}\left(\frac{\partial\phi}{\partial\xi}\right)^3$$

....

Thus, the balance equations may be formally expressed as

$$-c\Omega_\nu^{(\sigma)} = \int_0^T u_0^*(\zeta)B_\nu^{(\sigma)}(\zeta)d\zeta\,, \qquad (4.4.18)$$

$$-\lambda_l c_{l\nu}^{(\sigma)} = \int_0^T u_l^*(\zeta)B_\nu^{(\sigma)}(\zeta)d\zeta\,, \quad l \neq 0\,. \qquad (4.4.19)$$

Some lower-order quantities may be explicitly calculated. Clearly,

$$B_1(\zeta,[\phi]) = -2cD\frac{du_0}{d\zeta}\,. \tag{4.4.20}$$

Thus,

$$\begin{aligned}\Omega_1^{(1)} &= 2\left(0\left|D\frac{d}{d\zeta}\right|0\right) \equiv \mu\,,\\ c_{l1}^{(1)} &= 2c\lambda_l^{-1}\left(l\left|D\frac{d}{d\zeta}\right|0\right).\end{aligned} \tag{4.4.21}$$

After setting $\varepsilon = 1$, and replacing ξ by x, we get to the lowest approximation:

$$\frac{\partial\phi}{\partial t} = 1 + \mu\frac{\partial\phi}{\partial x}\,. \tag{4.4.22}$$

Here no dissipation or dispersion effects appear yet, and we proceed to the next order. After some calculations, we are led to the formulae

$$\Omega_2^{(1)} = (0|D|0) - 4\sum_{l\neq0}\lambda_l^{-1}\left(0\left|D\frac{d}{d\zeta}\right|l\right)\left(l\left|D\frac{d}{d\zeta}\right|0\right) \equiv \alpha\,, \tag{4.4.23 a}$$

$$\begin{aligned}\Omega_2^{(2)} = &-c\left(0\left|D\frac{d}{d\zeta}\right|0\right) - 4c\left(0\left|D\frac{d}{d\zeta}\right|0\right)\sum_{l\neq0}\lambda_l^{-1}\left(0\left|\frac{d}{d\zeta}\right|l\right)\left(l\left|D\frac{d}{d\zeta}\right|0\right)\\ &-4c\sum_{l,m\neq0}(\lambda_l\lambda_m)^{-1}(0|M|l,m)\left(l\left|D\frac{d}{d\zeta}\right|0\right)\left(m\left|D\frac{d}{d\zeta}\right|0\right)\\ &+4c\sum_{l\neq0}\lambda_l^{-1}\left(0\left|D\frac{d^2}{d\zeta^2}\right|l\right)\left(l\left|D\frac{d}{d\zeta}\right|0\right) \equiv \beta\,,\end{aligned} \tag{4.4.23 b}$$

where

$$(0|M|l,m) \equiv \int_0^{\bar{l}} u_0^*(\zeta)M(\zeta)u_l(\zeta)u_m(\zeta)\,d\zeta\,.$$

To this stage of approximation, (4.4.22) is modified to give

$$\frac{\partial\phi}{\partial t} = 1 + \mu\frac{\partial\phi}{\partial x} + \alpha\frac{\partial^2\phi}{\partial x^2} + \beta\left(\frac{\partial\phi}{\partial x}\right)^2. \tag{4.4.24}$$

Finally, the number density of the pulses is related to ϕ via

$$\begin{aligned}n(x,t) &= n_0 + \varrho(x,t)\,,\\ n_0 &= \bar{l}^{-1}\,,\\ \varrho(x,t) &= -c\bar{l}^{-1}\frac{\partial\phi}{\partial x}\,.\end{aligned}$$

Thus,

$$\frac{\partial \varrho}{\partial t} = \mu \frac{\partial \varrho}{\partial x} + \alpha \frac{\partial^2 \varrho}{\partial x^2} + 2\beta \varrho \frac{\partial \varrho}{\partial x}, \tag{4.4.25}$$

which coincides with the Burgers equation supplemented by a linear propagation term.

5. Mutual Entrainment

Collective oscillations in oscillator aggregates arise from the mutual entrainment among the constituent oscillators. This phenomenon seems to be of central importance in the self-organization in nature. The method of phase description I proves to be a convenient tool for approaching this problem.

5.1 Synchronization as a Mode of Self-Organization

Synchronization or *entrainment* is a key concept to the understanding of self-organization phenomena occurring in the fields of coupled oscillators of the dissipative type. We may even say that Part II is devoted to the consideration of this single mode of motion in various physical situations. Specifically, Chap. 6 is concerned with wave phenomena and pattern formation, which may be viewed as typical synchronization phenomena in distributed systems. In contrast, we shall study in Chap. 7 turbulence in reaction-diffusion systems, which is caused by *desynchronization* among local oscillators. Chapter 5 deals with self-synchronization phenomena in the discrete populations of oscillators where the way they are distributed in physical space is not important (for reasons stated later). We shall introduce some kind of randomness by assuming that the oscillators are either different in nature from each other or at best statistically identical. One may then expect phase-transition-like phenomena, characterized by the appearance or disappearance of collective oscillations in the oscillator communities. In describing such a new class of phase transitions, Method I turns out to be very useful.

It would be appropriate to mention briefly the relevance of such collective oscillations to real phenomena. Aside from oscillatory chemical reactions, systems which may be viewed as assemblies of limit cycle oscillators are abundand in nature. Such systems are, in fact, common in living organisms. The collective rhythms in living systems that arise from the cooperation of the individual cellular oscillators seem to play important roles in the coordination of life processes. For instance, the assembly of pacemaker cells in the human heart is considered to produce collective oscillations due to mutual entrainment, and this is of course vital to our life processes. Another important class of oscillatory phenomena shared by many living organisms is the circadian rhythms (Pavlidis, 1973; Bunning, 1973; Winfree, 1980). Under normal environmental conditions, such physiological oscillations are entrained by periodic external forces associated with the sunset-sunrise cycles. There exist a number of experimental observations indicating that spontaneous physiological oscillations can often persist (with a slight deviation in period from 24 hours) even in the complete absence of external driving forces. This naturally leads to the idea that some autonomous physiological oscillators of the limit cycle type are built into each living organism. If we go to an even more microscopic level, such physiological clocks may possibly prove to be a form of collective oscillations resulting from the cooperative interactions among cellular oscillators. More examples of multi-oscillator systems in living organisms and detailed accounts may be found in Winfree's book (1980).

The theoretical understanding of the origin of collective rhythmicity may best be obtained by studying its onset, that is, by treating it as a kind of phase transition or a bifurcation. In many biological oscillations including the examples mentioned above, the oscillators constituting an aggregate or a tissue are considered to be more or less similar to each other, although they could not be strictly identical. Specifically, their natural frequencies may be distributed over a certain range. Even if the frequencies are essentially identical in nature, they cannot be perfectly free from environmental fluctuations, so that they would at best be statistically identical. In any case, such randomness factors are destructive to mutual entrainment or to the formation of coherent rhythmicity. In contrast, coupling among the oscillators usually favors mutual synchronization. (Counter-examples exist, however, as is seen in Sect. 5.2.) The conflicts between such opposing tendencies are, in fact, common to all kinds of phase transitions.

If the constituent oscillators are identical, and no external noise is present, then infinitesimal mutual coupling would be sufficient to cause perfectly ordered behavior. More generally, one may imagine circumstances under which the phase-randomizing effects due to external noise or the distribution of natural frequencies are weak, and at the same time the mutual coupling is equally weak. This represents a non-trivial extreme case for which some transition phenomena between ordered and disordered motions may still be expected. As a system similar to this, one may think of a system of weakly coupled magnetic spins which shows a phase transition at low temperatures. A technical advantage in considering the above extreme situation is that one may then employ the simple perturbation method developed in Chap. 3 (possibly with some suitable modification or generalization). Mathematically, a new feature arises when the perturbation εp comes to include rapidly fluctuating external forces. This case will be treated in Sect. 5.3 which represents the only part of this book where we come a little into contact with statistical theories.

In spite of its great potential importance, the subject of mutual entrainment in multi-oscillator systems has been little explored so far. The earlier attempt due to Winfree is based on a phase description (Winfree, 1967). There have been some efforts to make his idea more precise in some respects (Kuramoto, 1975, 1981). Approaches not based on a phase description also exist (Aizawa, 1976; Yamaguchi et al., 1981). The theory presented below is partly in common with Neu's recent work on the populations of oscillators, although the latter does not treat phase-transition-like phenomena (Neu, 1980). There are some phase description approaches on the mutual entrainment of *two* oscillators (Neu, 1979b; Fujii and Sawada, 1978). This work was partly motivated by Marek an Stuchl's experiment on interacting oscillatory cells in the Belousov-Zhabotinsky reaction (Marek and Stuchl, 1975).

Finally, we note that in this chapter, especially in Sects. 5.5 – 7, we illustrate how the slaving principle works even when fluctuating forces are present. In this connection, the detailed treatment of the slaving principle including noise, which was recently presented by Haken and Wunderlin (1982), is worth mentioning, although they treat the problem in a physical or mathematical context rather different from the present one.

5.2 Phase Description of Entrainment

By studying step by step cases (a) – (c), arranged in order of increasing complexity below, one may realize how Method I is suited for treating various synchronization phenomena. Before going into specific arguments, however, let us generalize the class of the perturbation $\varepsilon \boldsymbol{p}$ in (3.2.2) to include time dependence:

$$\frac{dX}{dt} = F(X) + \varepsilon p(X, t) . \tag{5.2.1}$$

The lowest-order results (3.2.8, 9) are then modified as

$$\frac{d\phi}{dt} = 1 + \varepsilon \Omega(\phi, t) , \quad \text{where} \tag{5.2.2}$$

$$\Omega(\phi, t) = Z(\phi) \cdot \Pi(\phi, t) , \quad \text{with} \tag{5.2.3}$$

$$\Pi(\phi, t) \equiv p(X_0(\phi), t) .$$

Up to this stage, the generalization is trivial. When one averages with respect to rapidly oscillating processes, however, one must be very careful, and this point turns out to be crucial to the phase description of synchronization.

5.2.1 One Oscillator Subject to Periodic Force

Let the perturbation $\varepsilon \boldsymbol{p}$ represent a weak periodic force of period T', while the period of the native oscillator is denoted by T, i.e.,

$$\Omega(\phi, t) = \Omega(\phi + T, t) = \Omega(\phi, t + T') . \tag{5.2.4}$$

It is assumed that the difference $T - T'$ is small and is of order ε; we put

$$1 - \frac{T}{T'} = \varepsilon \Delta . \tag{5.2.5}$$

It is expected that the oscillator comes to oscillate with exactly the same frequency as the external one if $|\Delta|$ is below some critical value Δ_c. More general entrainment in which some integer multiples of T and T' become identical could be treated in a similar manner by assuming

$$1-\frac{mT}{nT'}=O(\varepsilon) \qquad m, n \text{ integers}, \tag{5.2.6}$$

although this case will not be considered here.

Let ψ denote a new phase variable defined by

$$\phi=\frac{T}{T'}t+\psi . \tag{5.2.7}$$

Note that the above ψ is not the same as the quantity we have so far worked with under the same notation. It is obvious that if ψ is constant in time, this implies that the oscillator is entrained to the external periodicity. The equation for ψ now takes the form

$$\frac{d\psi}{dt}=\varepsilon\left[\Delta+\Omega\left(\frac{T}{T'}t+\psi,t\right)\right]. \tag{5.2.8}$$

This shows that ψ is a slowly varying function of t. On the other hand, the quantity Ω, if viewed as a function of t and ψ, is T'-periodic in t (and T-periodic in ψ). Since the slow variable ψ would hardly change during the period T', one may safey time-average Ω over this interval with ψ kept constant. In this way, we obtain (after setting $\varepsilon=1$)

$$\frac{d\psi}{dt}=\Delta+\Gamma(\psi), \qquad \text{where} \tag{5.2.9}$$

$$\Gamma(\psi)=\Gamma(\psi+T)=\frac{1}{T'}\int_0^{T'}\Omega\left(\frac{T}{T'}t+\psi,t\right)dt .$$

Entrainment to the external periodicity occurs if (5.2.9) has a stable-equilibrium solution. In that case, we have at least one stable-unstable pair of equilibria (denoted as $\psi_0^{(s)}$ and $\psi_0^{(u)}$, respectively) in the interval $(0, T)$ (Fig. 5.1 a). A given equilibrium solution ψ_0 is stable if $d\Gamma/d\psi|_{\psi_0}$ is negative, and unstable if it is positive. It may of course happen that more stable-unstable pairs of equilibrium points appear (Fig. 5.1 b). When the condition for entrainment is not satisfied (Fig. 5.1 c), the oscillator gains a frequency different from the external one. Let this frequency difference be $\Delta\omega$, or

$$\Delta\omega=\frac{2\pi}{T}\overline{\frac{d\phi}{dt}}-\frac{2\pi}{T'}=\frac{2\pi}{T}\overline{\frac{d\psi}{dt}}, \tag{5.2.10}$$

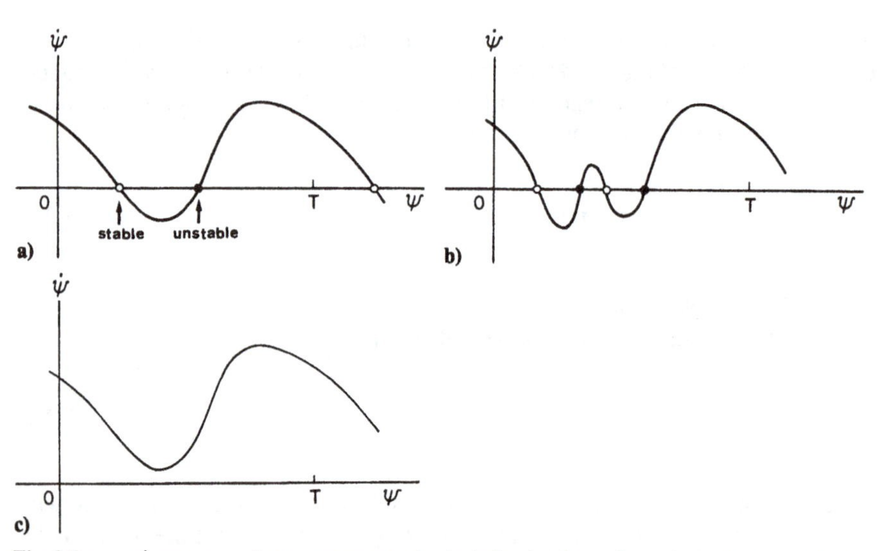

Fig. 5.1 a – c. $\dot{\psi}$ versus ψ, where ψ represents the deviation in phase of a periodically forced limit cycle oscillator. The oscillator may be entrained (**a** or **b**), or not entrained (**c**)

where the bar indicates the long-time average. One may calculate $\Delta\omega$ as follows. First, (5.2.9) is integrated to give

$$\int_0^{\psi} \frac{dx}{\Delta + \Gamma(x)} = h\psi + f(\psi) = t - t_0 ,$$

where

$$h = \frac{1}{T} \int_0^{T} \frac{dx}{\Delta + \Gamma(x)} ,$$

and $f(\psi)$ is a certain T-periodic function of ψ. Thus, ψ varies at the average rate h^{-1}, and we get from (5.2.10)

$$\Delta\omega = 2\pi/hT .$$

In particular, when Δ (supposed to be positive, for simplicity) is very close to the critical value Δ_c, h is evaluated as

$$h \simeq \frac{1}{T} \int_0^{T} \frac{dx}{\Delta - \Delta_c + \alpha(x-x_0)^2} \simeq \frac{1}{T} \int_{-\infty}^{\infty} \frac{dx}{\Delta - \Delta_c + \alpha x^2} = \frac{\pi}{T\sqrt{\alpha(\Delta - \Delta_c)}} , \quad (5.2.11)$$

where α is the curvature of $\Gamma(x)$ at its minimum point x_0. Thus $\Delta\omega$ decreases as $\sqrt{\Delta - \Delta_c}$ when $\Delta \to \Delta_c$.

5.2.2 A Pair of Oscillators with Different Frequencies

Let us next consider the mutual entrainment between two oscillators which are symmetrically coupled to each other. Let a system of this kind be described by

$$\frac{dX_1}{dt} = F_1(X_1) + \varepsilon V(X_1, X_2) ,$$
$$\frac{dX_2}{dt} = F_2(X_2) + \varepsilon V(X_2, X_1) . \tag{5.2.12}$$

It is assumed that the oscillators, if uncoupled, are slightly different in nature from each other, the difference being characterized by the smallness parameter ε. Thus one may put

$$F_\alpha(X_\alpha) = F(X_\alpha) + \varepsilon f_\alpha(X_\alpha) , \quad \alpha = 1, 2 . \tag{5.2.13}$$

This implies that the difference in natural frequency is also of order ε. Thus the problem becomes very similar to that of (a), and a critical condition for entrainment is expected to exist.

Let (5.2.12) be expressed in the form

$$\frac{dX_\alpha}{dt} = F(X_\alpha) + \varepsilon p_\alpha ,$$
$$p_\alpha = V(X_\alpha, X_{\alpha'}) + f_\alpha(X_\alpha) , \quad \alpha = 1, 2, \alpha \neq \alpha' . \tag{5.2.14}$$

The period of the unperturbed oscillator, which is actually an imaginary oscillator, is denoted by T. It is straightforward to derive coupled equations for ϕ_1 and ϕ_2 (the phases of the oscillators we are considering). They are

$$\frac{d\phi_1}{dt} = 1 + \varepsilon[Z(\phi_1) V(\phi_1, \phi_2) + g_1(\phi_1)] ,$$
$$\frac{d\phi_2}{dt} = 1 + \varepsilon[Z(\phi_2) V(\phi_2, \phi_1) + g_2(\phi_2)] , \quad \text{where} \tag{5.2.15}$$

$$g_\alpha(\phi_\alpha) = Z(\phi_\alpha) f_\alpha(X_0(\phi_\alpha)) \tag{5.2.16}$$

and $V(\phi_\alpha, \phi_{\alpha'})$ is the abbreviation of $V(X_0(\phi_\alpha), X_0(\phi_{\alpha'}))$. Note that $V(\phi_\alpha, \phi_{\alpha'})$ is T-periodic both in ϕ_α and $\phi_{\alpha'}$. Let ψ_α be defined as $\phi_\alpha = t + \psi_\alpha$. After time-averaging (5.2.15) over the period T for constant ψ_α, and putting ε equal to 1, we have

$$\frac{d\psi_1}{dt} = \Gamma(\psi_1 - \psi_2) + \omega_1 ,$$
$$\frac{d\psi_2}{dt} = \Gamma(\psi_2 - \psi_1) + \omega_2 , \tag{5.2.17a}$$

where

$$\Gamma(\psi_\alpha - \psi_{\alpha'}) = \frac{1}{T}\int_0^T Z(t+\psi_\alpha)\,V(t+\psi_\alpha, t+\psi_{\alpha'})\,dt\,,$$
$$\omega_\alpha = \frac{1}{T}\int_0^T g_\alpha(t+\psi_\alpha)\,dt\,. \tag{5.2.17b}$$

The condition for mutual entrainment is more easily seen by rewriting (5.2.17a) in terms of the phase difference $\psi \equiv \psi_1 - \psi_2$ as

$$\frac{d\psi}{dt} = \Delta + B(\psi)\,, \tag{5.2.18}$$

where $B(\psi) = \Gamma(\psi) - \Gamma(-\psi)$ and $\Delta = \omega_1 - \omega_2$. Since B is an odd function of ψ and T-periodic, the values of ψ given by

$$\psi_0 = \frac{nT}{2}\,, \qquad n = \text{integer}\,,$$

are the zeros of B, although there may be more zeros of B. If the two oscillators are identical (i.e., $\Delta = 0$), then the above values of ψ_0 give some equilibrium points of (5.2.18). The stability of these equilibrium points depends entirely on the sign of $dB/d\psi|_0$, or the sign of $d\Gamma/d\psi|_0$. If $d\Gamma/d\psi|_0 < 0$, stable equilibria are given by values of ψ_0 with even n, while if $d\Gamma/d\psi|_0 > 0$, those values with odd n are stable. In the former case, the phases of the oscillators become identical after entrainment, while in the latter case, they are spaced apart from each other by a half cycle $T/2$. Hence the mutual coupling may be called *attractive* if $d\Gamma/d\psi|_0 < 0$, and *repulsive* if $d\Gamma/d\psi|_0 > 0$.

5.2.3 Many Oscillators with Frequency Distribution

Formally, this is a simple generalization of case (b) by changing the number of oscillators from 2 to N. Let the system be described by

$$\frac{dX_\alpha}{dt} = F(X_\alpha) + \varepsilon p_\alpha\,, \qquad \alpha = 1, 2, \ldots N\,, \tag{5.2.19}$$

where

$$p_\alpha = \sum_{\alpha' \neq \alpha} V_{\alpha\alpha'}(X_\alpha, X_{\alpha'}) + f_\alpha(X_\alpha)\,.$$

Analogously to (5.2.15), we have

$$\frac{d\phi_\alpha}{dt} = 1 + \varepsilon\left[\sum_{\alpha'} Z(\phi_\alpha)\,V_{\alpha\alpha'}(\phi_\alpha, \phi_{\alpha'}) + g_\alpha(\phi_\alpha)\right]. \tag{5.2.20}$$

By putting $\phi_\alpha = t + \psi_\alpha$, this is reduced to

$$\frac{d\psi_\alpha}{dt} = \sum_{\alpha'} \Gamma_{\alpha\alpha'}(\psi_\alpha - \psi_{\alpha'}) + \omega_\alpha , \tag{5.2.21a}$$

where

$$\Gamma_{\alpha\alpha'}(\psi_\alpha - \psi_{\alpha'}) = \frac{1}{T}\int_0^T Z(t+\psi_\alpha) V_{\alpha\alpha'}(t+\psi_\alpha, t+\psi_{\alpha'})\,dt \tag{5.2.21b}$$

and

$$\omega_\alpha = \frac{1}{T}\int_0^T g_\alpha(t+\psi_\alpha)\,dt .$$

Unfortunately, it is generally impossible to treat (5.2.21) analytically. We therefore consider in Sect. 5.4 a soluble model where the coupling $\Gamma_{\alpha\alpha'}$ and the distribution of the natural frequency ω_α are given by some simple functions. An interesting feature revealed, then, is that a critical condition for the appearance of collective oscillations exists. Before going into such discussions, however, we make a few remarks on a simple coupled-oscillator model for which Γ may be calculated explicitly.

5.3 Calculation of Γ for a Simple Model

Consider a pair of "Ginzburg-Landau oscillators" which are interacting through a discretized version of diffusion. The evolution equations for the complex amplitudes $W = X + \mathrm{i}Y$ and $W' = X' + \mathrm{i}Y'$ are then given by

$$\frac{dW}{dt} = (1 + \mathrm{i}c_0)W - (1 + \mathrm{i}c_2)|W|^2 W + \varepsilon(1 + \mathrm{i}c_1)(W' - W) \tag{5.3.1}$$

and a similar equation for W'. For this model, $\Gamma(\psi - \psi')$ may be calculated explicitly. In terms of the vectors $\boldsymbol{U} \equiv (X, Y)$ and $\boldsymbol{U}' \equiv (X', Y')$, (5.3.1) may be expressed as

$$\frac{d\boldsymbol{U}}{dt} = \boldsymbol{F}(\boldsymbol{U}) + \varepsilon \boldsymbol{V}(\boldsymbol{U}, \boldsymbol{U}') , \quad \text{where} \tag{5.3.2}$$

$$\boldsymbol{F}(\boldsymbol{U}) = \begin{pmatrix} X - c_0 Y - (X - c_2 Y)(X^2 + Y^2) \\ Y + c_0 X - (Y + c_2 X)(X^2 + Y^2) \end{pmatrix} \quad \text{and} \tag{5.3.3}$$

$$\boldsymbol{V}(\boldsymbol{U}, \boldsymbol{U}') = \begin{pmatrix} X' - X - c_1(Y' - Y) \\ Y' - Y + c_1(X' - X) \end{pmatrix} . \tag{5.3.4}$$

The only quantities we need are u_0^*, $S(\phi)$ and $X_0(\phi)$ [i.e., the periodic solution of $dU/d\phi = F(U)$], since they determine $Z(\phi)$ [through (3.4.9)] and $V(\phi, \phi')$, and hence $\Gamma(\psi - \psi')$ [through (5.2.21b)]. But we have already calculated u_0^* in (3.5.12b), $S(\phi)$ in (3.5.10) and $X_0(\phi)$ in (3.5.5). From these expressions, one immediately gets

$$V(\phi, \phi') = \begin{pmatrix} \cos\Theta' - \cos\Theta - c_1(\sin\Theta' - \sin\Theta) \\ c_1(\cos\Theta' - \cos\Theta) + \sin\Theta' - \sin\Theta \end{pmatrix}, \tag{5.3.5}$$

where $\Theta \equiv \omega_0\phi$. It is easily to confirm that the dependence of $Z(\phi)\, V(\phi, \phi')$ on ϕ and ϕ' is only through $\phi - \phi'$ for our simple model, so that there is no need to time average. In fact, we have

$$\begin{aligned} Z(\phi)\, V(\phi, \phi') &= -\omega_0^{-1}((1 + c_1c_2)\sin[\omega_0(\psi - \psi')] \\ &\quad + (c_1 - c_2)\{1 - \cos[\omega_0(\psi - \psi')]\}) \\ &= \Gamma(\psi - \psi'). \end{aligned} \tag{5.3.6}$$

It is interesting to examine whether the pair coupling in the model is attractive or repulsive. We saw that the coupling is attractive if $d\Gamma/d\phi|_0 < 0$, and repulsive if $d\Gamma/d\psi|_0 > 0$. But (5.3.6) shows that

$$\left.\frac{d\Gamma}{d\psi}\right|_0 = -(1 + c_1c_2), \tag{5.3.7}$$

so that the coupling is either attractive or repulsive depending on the parameter values. The appearance of the term $1 + c_1c_2$ is worth noting. The same quantity appeared in Sect. 3.5 as a coefficient of the nonlinear phase diffusion equation (3.5.14). It is now seen that the repulsive coupling in this discretized Ginzburg-Landau model corresponds precisely to the negative phase diffusion in the continuum limit of the same model. A consequence of the negative diffusion is chemical turbulence as discussed in Chap. 7.

5.4 Soluble Many-Oscillator Model Showing Synchronization-Desynchronization Transitions

The population models of limit cycle oscillators which we obtained in Sect. 5.2 (c) seem to have been seldom investigated in the past. Although a general analytical treatment of (5.2.21) would be difficult, there certainly exists, in the limit of large N, a special subclass of systems for which a number of interesting analytical results are available.

It is natural to expect that the simplest analytical approach to the N-body cooperative system (5.2.21) would be provided by something similar to the mean field idea of thermodynamic phase transitions. In the present problem, the individual oscillators play the part of cooperative units like magnetic spins. It is a well-known fact that the mean field theory becomes more and more accurate as the effective number of spins coupled to each spin becomes larger. In this connection, there exists an idealized model called the Husimi-Temperly model in which each spin is postulated to interact with all the remaining spins with equal strength; for this model, the mean field theory is known to be exact. The same idea seems to be transferable to our oscillator community.

Let $\Gamma_{\alpha\alpha'}$ be identical for all pairs (α, α') and have a magnitude of order N^{-1}. This property ensures that the typical strength of the net local field experienced by each oscillator is independent of the total number N. We put

$$\Gamma_{\alpha\alpha'}(\psi_\alpha - \psi_{\alpha'}) = N^{-1}\Gamma(\psi_\alpha - \psi_{\alpha'}) . \tag{5.4.1}$$

Furthermore. let us take a simple trigonometric function for the T-periodic function Γ:

$$\Gamma(\psi_\alpha - \psi_{\alpha'}) = -K\sin(\psi_\alpha - \psi_{\alpha'}) , \tag{5.4.2}$$

where a suitable time scale has been adopted in which $T = 2\pi$. We assume K is positive, which makes the coupling attractive; otherwise no collective oscillations would be possible for the long-range interaction model due to the cancellation of the internal field. A more general model,

$$\Gamma(\psi_\alpha - \psi_{\alpha'}) = -K\sin(\psi_\alpha - \psi_{\alpha'} + \psi_0) , \tag{5.4.3}$$

with $0 \le |\psi_0| < \pi/2$, might also be interesting. However, the analysis would then become much more difficult, for the reasons stated later, and such a model will not be considered here. Let $g(\omega)$ denote the normalized number density of the oscillators whose natural frequencies ω_α coincide with ω. $g(\omega)$ is assumed to be symmetric about some frequency ω_0, or

$$g(\omega_0 + \omega) = g(\omega_0 - \omega) . \tag{5.4.4}$$

We now try to solve (5.2.21 a) or

$$\frac{d\psi_\alpha}{dt} = \omega_\alpha - N^{-1}K \sum_{\alpha'=1}^{N} \sin(\psi_\alpha - \psi_{\alpha'}) \tag{5.4.5}$$

under the condition (5.4.4).

Let $\hat{n}(\{\psi_\alpha\}; \psi)$ denote a dynamical variable representing the number density of the oscillators which have phase ψ, or

$$\hat{n}(\{\psi_\alpha\}; \psi) = N^{-1} \sum_{\alpha=1}^{N} \delta(\psi - \psi_\alpha) . \tag{5.4.6}$$

Since N is supposed to be infinitely large, one may expect that an infinitely large number of oscillators can fall into an arbitrarily small but finite interval $\Delta\psi$ about given ψ. One may then define an infinitesimally smoothened number density $n(\psi,t)$ through

$$n(\psi,t) = \lim_{\Delta\psi\to 0}\lim_{N\to\infty}(\Delta\psi)^{-1}\int_{\psi-(\Delta\psi/2)}^{\psi+(\Delta\psi/2)} \hat{n}(\{\psi_\alpha\};\psi')d\psi' , \tag{5.4.7}$$

and whenever $\hat{n}$ appears in the formulation, let us replace it by n. Note that $n(\psi,t)$ is 2π-periodic in ψ and satisfies

$$\int_0^{2\pi} n(\psi,t)d\psi = 1 . \tag{5.4.8}$$

We will be mainly concerned with the behavior of $n(\psi,t)$ rather than the individual solutions of (5.4.5). In fact, the former quantity is more relevant to macroscopic observations. Although the transient behavior of $n(\psi,t)$ would be difficult to study, one may say something about its long-time behavior. It is important to notice that (5.4.5) remains invariant under a uniform translation of all ψ_α,

$$\psi_\alpha \to \psi_\alpha + \psi_0 , \quad \alpha = 1,2,\ldots,N . \tag{5.4.9}$$

From this property it is expected that the simplest collective behavior may possibly be described by a uniform and stationary distribution of $n(\psi,t)$:

$$n(\psi,t) = n_0 = (2\pi)^{-1} , \tag{5.4.10}$$

whereas the second simplest collective behavior is considered to be a steadily traveling behavior of $n(\psi,t)$ in the cyclic ψ space:

$$n(\psi,t) = n(\psi - \Omega t) . \tag{5.4.11}$$

We expect that the latter type of solutions is associated with collective oscillations. However, we have not yet given a general definition of collective oscillations. It is quite natural to define collective oscillations as time-periodic motions of $\bar{X}(t)$, where $\bar{X}(t)$ is the simple average of $X_\alpha(t)$ over the entire system i.e.,

$$\bar{X}(t) = N^{-1}\sum_{\alpha=1}^{N} X_\alpha(t) .$$

In the lowest-order perturbation approximation which we adopted, $\bar{X}(t)$ is expressed as

$$\bar{X}(t) \simeq N^{-1}\sum_{\alpha=1}^{N} X_0(\phi_\alpha(t)) = \int_0^{2\pi} d\psi\, n(\psi,t)X_0(t+\psi) . \tag{5.4.12}$$

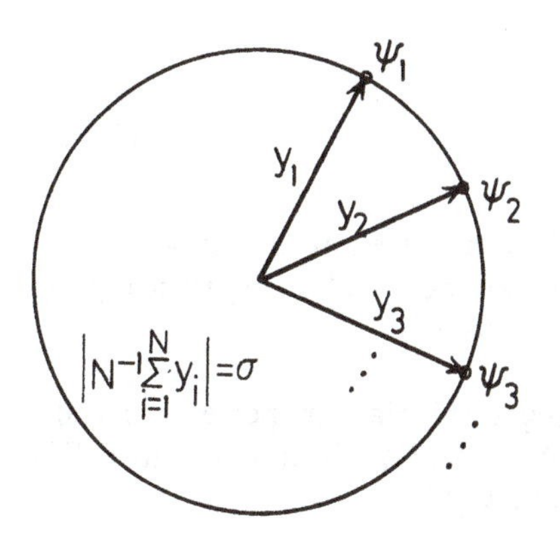

Fig. 5.2. Oscillators' phase states $\psi_1, \psi_2, \ldots, \psi_N$ distributed on a unit circle, and associated state vectors $y_1, y_2, \ldots, y_N$. The length of the vector sum $\sum y_i$ divided by N gives the magnitude of the order parameter

It is clear that constant n makes $\bar{X}$ time independent, hence no collective oscillations. In contrast, the traveling solutions (5.4.11) make $\bar{X}$ time periodic with frequency $1+\Omega$. Of course, there may exist various modes of motion as $t \to \infty$ other than (5.4.10, 11). In what follows, we shall, however, restrict our consideration to the above simplest possibilities; we want to know if particular classes of solutions such as (5.4.10) and (5.4.11) really exist, and if they do, how a transition between them is possible.

In analogy to the thermodynamic phase transitions, it is appropriate to define an order parameter. A convenient choice for this would be a complex quantity $\sigma \exp(\mathrm{i}\theta)$, defined by

$$\sigma e^{\mathrm{i}\theta} = \int_0^{2\pi} n(\psi') e^{\mathrm{i}\psi'} d\psi' . \tag{5.4.13}$$

Equivalently, one may distribute the N oscillators on a unit circle (Fig. 5.2), associate a vector of unit length to each oscillator, take a vector sum and divide it by N. Clearly, σ vanishes for the uniform distribution (5.4.10), and it is non-vanishing for the traveling density waves (5.4.11). From the definitions of n and the complex order parameter, (5.4.5) may be transformed into

$$\frac{d\psi_\alpha}{dt} = \omega_\alpha - K\sigma \sin(\psi_\alpha - \Omega t - \theta) . \tag{5.4.14}$$

Note that Ω and σ are the quantities yet to be determined, while θ may be chosen arbitrarily since it only fixes the initial phase of the traveling waves $n(\psi - \Omega t)$. Since Ω and σ were supposed to be time independent, one may integrate (5.4.14) to obtain ψ_α for each α. The solution set $\{\psi_\alpha\}$ obtained in this way determines $n(\psi, t)$, which gives σ via (5.4.13), and this σ must coincide with the same quantity in (5.4.14). In this way, we are led to a self-consistent equation for the order parameter.

Let Ψ_α denote the relative phase defined by

$$\Psi_\alpha = \psi_\alpha - \Omega t - \theta .$$

Then (5.4.14) becomes

$$\frac{d\Psi_\alpha}{dt} = \omega_\alpha - \Omega - K\sigma \sin \Psi_\alpha . \tag{5.4.15}$$

This equation is essentially the same as (5.2.9), and we already know that the solution shows distinct features depending on parameter values. Specifically, the oscillators fall into either of the following two groups.

(A) $|\omega_\alpha - \Omega| \leq K\sigma$. The oscillators belonging to this class are perfectly entrained to the oscillating internal field because (5.4.15) has equilibrium solutions. The stable equilibrium solution is given (in terms of ψ_α) by

$$\psi_\alpha = \Omega t + \theta + \sin^{-1}\left(\frac{\omega_\alpha - \Omega}{K\sigma}\right), \tag{5.4.16}$$

where the last term should be understood as the principal value, i.e., $|\sin^{-1} x| \leq \pi/2$. The frequency of the αth oscillator has now changed from ω_α to $\tilde{\omega}_\alpha$, where

$$\tilde{\omega}_\alpha = \Omega . \tag{5.4.17}$$

Clearly, the oscillators of this group form a synchronized cluster, which in turn generates an oscillating internal field in a self-consistent manner.

(B) $|\omega_\alpha - \Omega| > K\sigma$. The oscillators of this group fail to be entrained because their natural frequencies differ too much from the frequency of the internal field. Equation (5.4.15) is then integrated to give

$$\psi_\alpha = \tilde{\omega}_\alpha t + f((\tilde{\omega}_\alpha - \Omega)t), \tag{5.4.18}$$

where the modified frequency $\tilde{\omega}_\alpha$ of the αth oscillator is given by

$$\tilde{\omega}_\alpha = \Omega + (\omega_\alpha - \Omega)\sqrt{1 - \left(\frac{K\sigma}{\omega_\alpha - \Omega}\right)^2} \tag{5.4.19}$$

and $f(x)$ is a certain 2π-periodic function of x. It is seen that even if the oscillators fail to be entrained, their frequencies are modified due to the internal field. For the oscillators near the threshold of entrainment, we have $\tilde{\omega}_\alpha \simeq \Omega$, whereas for those far from the threshold, the $\tilde{\omega}_\alpha$ remain essentially unchanged from their natural frequencies.

Having thus obtained $\psi_\alpha(t)$ explicitly for each α, it is possible to construct the distribution $n(\psi, t)$ via $g(\omega)$. Corresponding to the two classes of solutions above, the density $n(\psi, t)$ may conveniently be divided into two parts, coming from the synchronized and desynchronized oscillators, or

$$n(\psi, t) = n_{\mathrm{s}}(\psi, t) + n_{\mathrm{ds}}(\psi, t) . \tag{5.4.20}$$

For the synchronized oscillators, the natural frequency ω_α may be viewed as a definite function of $\psi_\alpha - \Omega t$ given by (5.4.16). Thus the synchronized part n_s is related to g via

$$n_s(\psi, t) = n_s(\psi - \Omega t) = g(\omega(\psi - \Omega t)) \left| \frac{d\omega}{d(\psi - \Omega t)} \right| , \tag{5.4.21}$$

which gives

$$\begin{aligned} n_s(\psi - \Omega t) &= K\sigma g(\Omega + K\sigma \sin(\psi - \Omega t - \theta)) \cos(\psi - \Omega t - \theta) , \\ &\qquad \left(|\psi - \Omega t - \theta| \leq \frac{\pi}{2} \right) , \\ &= 0 , \qquad \left(|\psi - \Omega t - \theta| > \frac{\pi}{2} \right) . \end{aligned} \tag{5.4.22}$$

When the order parameter σ is non-vanishing, the desynchronized part $n_{\mathrm{ds}}(\psi, t)$ is expected to have a non-uniform distribution as a function of $\psi - \Omega t$. The origin of this non-uniformity may be seen from the following observation: the oscillators which are almost marginal (i.e., $\omega_\alpha - \Omega \simeq \pm K\sigma$) will find their phases staying for most of their period near $\psi_\alpha - \Omega t = \theta \pm \pi/2$ for which the velocity $|d(\psi - \Omega t)/dt|$ takes the minimum value. Thus the oscillators will be more concentrated near there than elsewhere. More generally, the probability $p(\psi - \Omega t, \omega_\alpha)$ that the phase value of ψ_α coincides with ψ at time t is considered to be inversely proportional to $|d(\psi_\alpha - \Omega t)/dt|_{\psi_\alpha = \psi}$, i.e.,

$$\begin{aligned} p(\psi - \Omega t, \omega_\alpha) &= \left| \frac{d(\psi_\alpha - \Omega t)}{dt} \right|^{-1}_{\psi_\alpha = \psi} \left(\int_0^{2\pi} d\psi_\alpha \left| \frac{d(\psi_\alpha - \Omega t)}{dt} \right|^{-1} \right)^{-1} \\ &= \frac{|\omega_\alpha - \Omega| \sqrt{1 - \left(\dfrac{K\sigma}{\omega_\alpha - \Omega} \right)^2}}{2\pi |\omega_\alpha - \Omega - K\sigma \sin(\psi - \Omega t - \theta)|} . \end{aligned} \tag{5.4.23}$$

Since the oscillators are not mutually correlated, apart from being subject to a common internal field, the density n_{ds} may simply be given by the superposition of $p(\psi - \Omega t, \omega_\alpha)$ over all desynchronized oscillators. Thus

$$\begin{aligned} n_{\mathrm{ds}}(\psi - \Omega t) &= \int_{|\omega - \Omega| > K\sigma} g(\omega) p(\psi - \Omega t, \omega) d\omega \\ &= \frac{1}{\pi} \int_{K\sigma}^{\infty} dx\, x g(\Omega + x) \frac{\sqrt{x^2 - (K\sigma)^2}}{x^2 - [K\sigma \sin(\psi - \Omega t - \theta)]^2} . \end{aligned} \tag{5.4.24}$$

We substitute the sum of (5.4.22) and (5.4.24) for $n(\psi)$ in (5.4.13), which leads to a self-consistent equation for σ. This equation becomes considerably simplified by virtue of the property

$$\int_0^{2\pi} n_{ds}(\psi') e^{i(\psi'-\theta)} d\psi' = 0 , \tag{5.4.25}$$

which follows from the equality $n_{ds}(\psi') = n_{ds}(\psi' + \pi)$. It should be noted, however, that (5.4.25) no longer holds for the model in (5.4.3) with nonvanishing ψ_0, and for this reason its analysis would be much more involved.

With the use of (5.4.22, 25), we have a self-consistent equation for σ in the form

$$\sigma = \int_0^{2\pi} n_s(\psi') e^{i(\psi'-\theta)} d\psi' = K\sigma \int_{-\pi/2}^{\pi/2} g(\Omega + K\sigma \sin x) \cos x \, e^{ix} dx . \tag{5.4.26}$$

Clearly, $\sigma = 0$ is always a solution. A non-zero solution is now sought. Equation (5.4.26) is then equivalent to a pair of equations

$$\int_{-\pi/2}^{\pi/2} dx \, g(\Omega + K\sigma \sin x) \cos x \sin x = 0 , \tag{5.4.27 a}$$

$$K \int_{-\pi/2}^{\pi/2} dx \, g(\Omega + K\sigma \sin x) \cos^2 x = 1 , \tag{5.4.27 b}$$

from which σ and Ω are to be determined. By assumption, $g(x)$ is symmetric about ω_0, so that (5.4.27 a) may be satisfied if and only if

$$\Omega = \omega_0 . \tag{5.4.28}$$

Equation (5.4.27 b) represents a transcendental equation for σ. We expect that a critical condition exists for the appearance of real σ. Near criticality, σ is expected to be small so that the left-hand side of (5.4.27 b) may be expanded in powers of σ:

$$1 - \frac{\pi}{2} K g(\omega_0) - \frac{\pi}{16} K^3 g^{(2)}(\omega_0) \sigma^2 + O(\sigma^3) = 0 , \tag{5.4.29}$$

where $g^{(2)}(\omega_0) \equiv d^2 g(\omega_0)/d\omega_0^2$. Thus the critical condition is given by

$$K = K_c \equiv \frac{2}{\pi g(\omega_0)} . \tag{5.4.30}$$

Let μ be a bifurcation parameter defined by

$$\mu = (K - K_c)/K_c .$$

One may expect that negative μ (i.e., weaker coupling) makes the zero solution stable, and positive μ (i.e., stronger coupling) unstable. Surprisingly enough, this seemingly obvious fact seems difficult to prove. The difficulty here comes from the fact that an infinitely large number of phase configurations $\{\psi_i; i=1,\ldots,N\}$ belong to an identical "macroscopic" state specified by a given value of σ. Near the point $\mu=0$, there exists a small-amplitude bifurcating solution

$$\sigma = \sqrt{\left|\frac{8g(\omega_0)\mu}{K_c^2 g^{(2)}(\omega_0)}\right|}\,. \tag{5.4.31}$$

The bifurcation is supercritical if $g^{(2)}(\omega_0)<0$, and subcritical if $g^{(2)}(\omega_0)>0$. It is expected that the supercritical bifurcating solution is stable, and the subcritical one is unstable. Again, this fact appears to be difficult to prove. Physically, why the sign of $g^{(2)}(\omega_0)$ determines the direction of branching of the non-zero solution may be qualitatively understood as follows. If g is concave at $\omega=\omega_0$, then the "nucleation" of synchronized oscillators, once initiated, will be speeded up due to the increasing number density of the oscillators just participating in the nucleation. As a result, the cluster growth will not be suppressed until its size reaches a fairly large value. This seems to be true even if the system is only infinitesimally above criticality. In contrast, for convex g at ω_0, the nucleation will slow down as it proceeds, so that the equilibrium size of the synchronized cluster will start from the zero value as K increases beyond K_c.

So far, the nucleation has been supposed to be initiated at the center of symmetry of g. This does not seem to be true, however, when g is concave there. The reason is that if $g^{(2)}(\omega_0)>0$ there must be some local maximum of g, say at ω_1, such that $g(\omega_1)>g(\omega_0)$. This implies the existence of a lower critical value $K_c'=2/\pi g(\omega_1)$ for the onset of nucleation which occurs at ω_1. As a consequence, the trivial solution, i.e., $\sigma=0$, can never persist up to K_c even as a metastable state, which makes a sharp contrast to the usual subcritical bifurcations. The minimum critical value of K for the onset of nucleation is thus given by $K_{c0}\equiv 2/\pi\,\mathrm{Max}\{g(\omega)\}$. The analysis of what happens for $K>K_{c0}$ would however be made difficult due to the asymmetry of g about its maximum point. Let us restrict ourselves to g symmetric about ω_0 as before, and let the maxima of g be situated at $\omega_\pm \equiv \omega_0 \pm \Delta$. Although the analysis would still be difficult, one may qualitatively predict what occurs as K is increased beyond K_{c0}. The clusters formed about ω_+ and ω_- will be small for K near K_{c0}, so that they will behave almost independently of each other. As their size becomes larger with K, they will come to behave like a coupled pair of giant oscillators, and for even stronger coupling they will eventually be entrained to each other to form a single giant oscillator (Fig. 5.3). For more general (asymmetric) g, too, a similar picture may hold qualitatively (Fig. 5.4).

We now come back to the simple case of g symmetric about its true maximum at ω_0. Then the fraction r of the population forming a synchronized cluster may be calculated for small cluster size, i.e., near criticality. We find

$$r=\frac{N_s}{N}=\int_{\omega_0-K\sigma}^{\omega_0+K\sigma} g(\omega)d\omega = 2K\sigma g(\omega_0)+O(\sigma^3)\,. \tag{5.4.32}$$

A quantity of particular interest is the distribution of the effective frequencies, which we denote by $G(\tilde{\omega})$. This is clearly related to $g(\omega)$ via

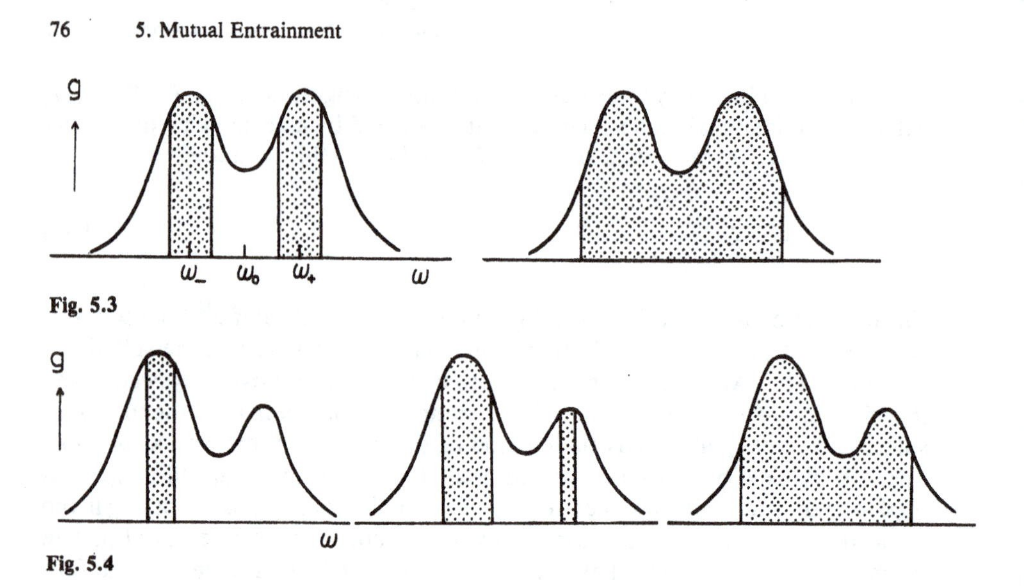

Fig. 5.3

Fig. 5.4

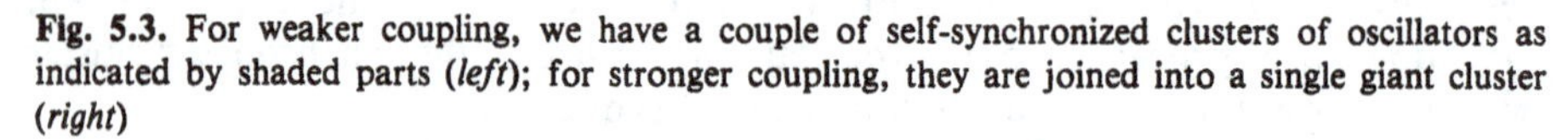

Fig. 5.3. For weaker coupling, we have a couple of self-synchronized clusters of oscillators as indicated by shaded parts (*left*); for stronger coupling, they are joined into a single giant cluster (*right*)

Fig. 5.4. Formation of self-synchronized oscillator clusters and their mutual synchronization in the case of an asymmetric distribution of the natural frequencies. Compare with Fig. 5.3

$$G(\tilde{\omega}) = g(\omega)\left|\frac{d\omega}{d\tilde{\omega}}\right| . \tag{5.4.33}$$

As we did for $n(\psi, t)$ in (5.4.20), we may conveniently express G as a sum of synchronized and desynchronized parts:

$$G(\tilde{\omega}) = G_s(\tilde{\omega}) + G_{ds}(\tilde{\omega}) . \tag{5.4.34}$$

The synchronized part is of course concentrated on frequency ω_0, i.e.,

$$G_s(\tilde{\omega}) = r\delta(\tilde{\omega} - \omega_0) . \tag{5.4.35}$$

The desynchronized part is calculated through the expression for $\tilde{\omega}$ in (5.4.19), and we have

$$G_{ds}(\tilde{\omega}) = g\left(\omega_0 + \sqrt{(\tilde{\omega} - \omega_0)^2 + (K\sigma)^2}\right) \frac{|\tilde{\omega} - \omega_0|}{\sqrt{(\tilde{\omega} - \omega_0)^2 + (K\sigma)^2}} . \tag{5.4.36}$$

In the absence of collective oscillations (i.e., $\sigma = 0$), $G_{ds}(x)$ is identical to $g(x)$, while for non-vanishing σ, $G_{ds}(x)$ has interesting properties:

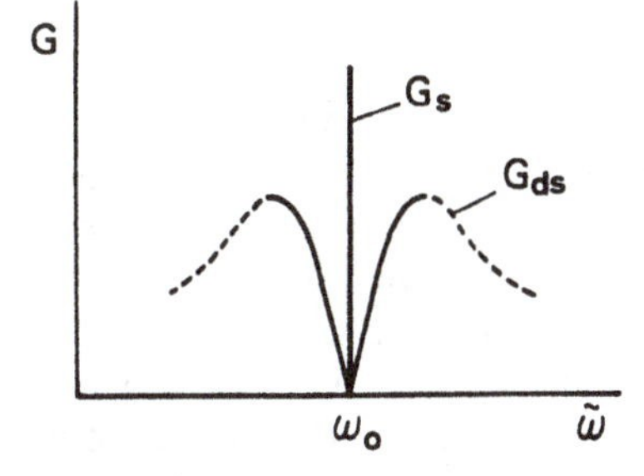

Fig. 5.5. Sharp population drop near the frequency of synchronization

$$G_{\mathrm{ds}}(\omega_0)=0\,,\qquad \lim_{x\to\omega_0\pm}\frac{dG_{\mathrm{ds}}}{dx}=\pm\frac{g(\omega_0+K\sigma)}{K\sigma}\,. \tag{5.4.37}$$

These properties give a peculiar look to G as shown in Fig. 5.5. Physically, this seems to reflect the fact that the oscillators whose natural frequencies are close to ω_0 are pulled perfectly into the central frequency to form a sharp peak, while this results in a remarkable population decrease around this peak. This is the very feature that caught N. Wiener's attention (1965), in connection with the spectrum of the α rhythm of the human brain waves.

Finally, we note that various quantities can be calculated explicitly over the entire parameter range if $g(\omega)$ is a Lorentzian, or,

$$g(\omega)=\frac{\gamma}{\pi}\,\frac{1}{(\omega-\omega_0)^2+\gamma^2}\,.$$

Then the essential parameter is

$$\eta\equiv 2\gamma/K\,.$$

We only show some of the final results since their calculations are straightforward:

$$\begin{aligned}\sigma&=\sqrt{1-\eta}\,,\qquad \eta\le 1\,,\\ &=0\,,\qquad\qquad \eta>1\,;\end{aligned} \tag{5.4.38}$$

$$\begin{aligned}r&=\frac{2}{\pi}\tan^{-1}\frac{2\sqrt{1-\eta}}{\eta}\,,\qquad \eta\le 1\,,\\ &=0\,,\qquad\qquad \eta>1\,;\end{aligned} \tag{5.4.39}$$

$$G(\tilde\omega)=r\delta(\tilde\omega-\omega_0)+\frac{\gamma}{\pi}\,\frac{|\tilde\omega-\omega_0|}{[(\tilde\omega-\omega_0)^2+(K\sigma)^2+\gamma^2]\sqrt{(\tilde\omega-\omega_0)^2+(K\sigma)^2}}\,, \tag{5.4.40}$$

$$n(\psi)=\frac{\gamma}{2\pi}\,\frac{\sqrt{(K\sigma)^2+\gamma^2}+K\sigma\cos\psi}{(K\sigma\sin\psi)^2+\gamma^2}\,. \tag{5.4.41}$$

5.5 Oscillators Subject to Fluctuating Forces

The population composed of identical oscillators coupled through an attractive interaction has been seen to behave like a perfectly self-entrained giant oscillator. A possible way to cause synchronization-desynchronization transitions is to introduce some distribution in natural frequencies, which we did in the previous section. If the oscillators are only weakly coupled (otherwise the phase description would not work), the width of the frequency distribution must also be small, so that the corresponding ordering and disordering forces may be counterbalanced. As was pointed out in Sect. 5.1, the situation is then quite similar to the thermodynamic phase transitions occurring at low temperatures, for which weak thermal agitations present form the counterpart of the frequency distribution of small width. This analogy naturally leads to the idea that synchronization-desynchronization transitions in populations might occur as well by introducing, in place a frequency distribution, some random forces acting on the individual oscillators; otherwise the oscillators may be identical in nature. No technical problems arise when stochastic forces are included in $\varepsilon \boldsymbol{p}$, except that one must be careful about time averaging as described later. The fluctuating forces must be assumed weak so that they may counterbalance the weak cooperative forces. Whether the previous deterministic model or the present stochastic one is more appropriate depends on the specific physical problems concerned. Stochastic models seem, however, to possess at least the advantage of being mathematically simpler. In addition, adopting stochastic models causes no difficulties associated with the stability of various solutions, whereas difficulties of this kind were seen to be unavoidable for the previous deterministic model. Analogously to the discussion in Sect. 5.2, we shall begin with a one-oscillator problem, and then proceed to two- and many-oscillator problems.

5.5.1 One Oscillator Subject to Stochastic Forces

Let a limit cycle oscillator be exposed to some weak random forces which may depend on the state variable $\boldsymbol{X}$. The governing equation is a nonlinear stochastic equation:

$$\frac{d\boldsymbol{X}}{dt} = \boldsymbol{F}(\boldsymbol{X}) + \varepsilon \boldsymbol{f}(\boldsymbol{X}, t) . \tag{5.5.1}$$

Interpreting the random forces as a perturbation $\varepsilon \boldsymbol{p}$ in (5.2.1), one may immediately write down the equation for ϕ, to the lowest-order approximation, as

$$\frac{d\phi}{dt} = 1 + \varepsilon g(\phi, t) , \quad \text{where} \tag{5.5.2}$$

$$g(\phi, t) = \boldsymbol{Z}(\phi) \cdot \boldsymbol{f}(\boldsymbol{X}_0(\phi), t) .$$

Thus $g(\phi, t)$ is a T-periodic function of ϕ but stochastically dependent on t unlike the corresponding quantity $\Omega(\phi, t)$ in (5.2.2). In the usual stochastic models, the random forces are supposed to change very rapidly about the zero value, and this we also assume for g. As a consequence, a naive application of the time-averaging procedures we employed previously leads to trivial results. Instead, one should first try to transform (5.5.2) into an evolution equation for the probability distribution $P(\phi, t)$, which is in fact possible with suitable assumptions for g, and then one may safely take a time average. As is most commonly done, we let the random forces g be Gaussian and delta-correlated with the vanishing mean value. Let $\langle \cdots \rangle$ denote a statistical average. Then,

$$\langle g(\phi, t) \rangle = 0\,, \tag{5.5.3a}$$

$$\langle g(\phi, t)\, g(\phi, t') \rangle = 2D(\phi)\, \delta(t-t')\,. \tag{5.5.3b}$$

It is a well known fact (Stratonovich, 1967) that an equation of the form (5.5.2) is then equivalent to the Fokker-Planck equation

$$\frac{\partial P}{\partial t} = -\frac{\partial I}{\partial \phi}, \quad \text{where} \tag{5.5.4}$$

$$I = \left(1 + \frac{\varepsilon}{2}\frac{dD}{d\phi}\right) P - \varepsilon \frac{\partial}{\partial \phi}(DP)\,.$$

Rigorously speaking, the above ε should be read ε^2, but we retain the above form and simply interpret ε as some small parameter.

The change of variable from ϕ to $\psi \equiv \phi - t$ is now made. Let $Q(\psi, t)$ denote the probability distribution for ψ, or

$$Q(\psi, t) = P(t + \psi, t)\,.$$

Then it immediately follows that

$$\frac{\partial Q}{\partial t} = -\varepsilon \frac{\partial J}{\partial \psi}, \quad \text{where} \tag{5.5.5}$$

$$J = \frac{1}{2}\frac{\partial D(t+\psi)}{\partial \psi} Q - \frac{\partial}{\partial \psi}[D(t+\psi)Q]\,.$$

Clearly, $Q(\psi, t)$ is a slowly varying function of t. Compared to $Q(\psi, t)$, the quantities $\partial D/\partial \psi$ and D fluctuate much faster, so that the latter may safely be time averaged over the period T with constant Q. Since the average of $\partial D/\partial \psi$ is zero, one is left (after putting ε back equal to 1) with a simple diffusion equation,

$$\frac{\partial Q}{\partial t} = \bar{D}\frac{\partial^2 Q}{\partial \psi^2}, \quad \text{where} \tag{5.5.6}$$

$$\bar{D} = \frac{1}{T}\int_0^T D(t+\psi)\,dt\,.$$

This kind of dephasing on limit cycle orbits was first discussed by Tomita and Tomita by means of a specific birth-death model of chemical reactions (Tomita and Tomita, 1974).

5.5.2 A Pair of Oscillators Subject to Stochastic Forces

Suppose that a pair of oscillators, which are identical in nature, are weakly coupled to each other and under the influence of some external stochastic forces. The following model would then be appropriate:

$$\frac{dX_\alpha}{dt} = F(X_\alpha) + \varepsilon p_\alpha\,, \qquad \alpha = 1, 2\,. \tag{5.5.7}$$

Here

$$p_\alpha = V(X_\alpha, X_{\alpha'}) + f_\alpha(X_\alpha, t)\,, \qquad \alpha \neq \alpha'\,.$$

The lowest-order perturbation theory applied to (5.5.7) yields

$$\frac{d\phi_\alpha}{dt} = 1 + \varepsilon[Z(\phi_\alpha)V(\phi_\alpha, \phi_{\alpha'}) + g_\alpha(\phi_\alpha, t)]\,, \quad \text{where} \tag{5.5.8}$$

$$g_\alpha(\phi_\alpha, t) = Z(\phi_\alpha)\cdot f_\alpha(X_0(\phi_\alpha, t))\,,$$

and $V(\phi_\alpha, \phi_{\alpha'})$ is the abbreviation of $V(X_0(\phi_\alpha), X_0(\phi_{\alpha'}))$. The random forces g_α are again assumed to be Gaussian and delta-correlated with the zero mean value:

$$\langle g_\alpha(\phi_\alpha, t)\rangle = 0\,, \tag{5.5.9a}$$

$$\langle g_\alpha(\phi_\alpha, t)\, g_{\alpha'}(\phi_{\alpha'}, t')\rangle = 2\delta_{\alpha\alpha'}D(\phi_\alpha)\,\delta(t-t')\,. \tag{5.5.9b}$$

The joint probability distribution $P(\phi_1, \phi_2, t)$ is then governed by

$$\frac{\partial P}{\partial t} = -\left(\frac{\partial I_1}{\partial \phi_1} + \frac{\partial I_2}{\partial \phi_2}\right), \quad \text{where} \tag{5.5.10}$$

$$I_\alpha = \left\{1 + \varepsilon\left[Z(\phi_\alpha)\cdot V(\phi_\alpha, \phi_{\alpha'}) + \frac{1}{2}\,\frac{dD(\phi_\alpha)}{d\phi_\alpha}\right]\right\}P - \varepsilon\frac{\partial}{\partial \phi_\alpha}[D(\phi_\alpha)P]\,.$$

Let ψ_α be defined by $\phi_\alpha = 1 + \psi_\alpha$, and let $Q(\psi_1, \psi_2, t)$ denote the joint probability for ψ_1 and ψ_2. It is straightforward to obtain

$$\frac{\partial Q}{\partial t} = -\varepsilon\left(\frac{\partial J_1}{\partial \phi_1} + \frac{\partial J_2}{\partial \phi_2}\right), \quad \text{where} \tag{5.5.11}$$

$$J_\alpha = \left[Z(t+\psi_\alpha)\cdot V(t+\psi_\alpha, t+\psi_{\alpha'}) + \frac{1}{2}\,\frac{\partial D(t+\psi_\alpha)}{\partial \psi_\alpha}\right]Q - \frac{\partial}{\partial \psi_\alpha}[D(t+\psi_\alpha)Q], \quad \alpha \neq \alpha'.$$

Rapidly oscillating quantities in J_α are now time averaged. After putting ε equal to 1, we obtain

$$\frac{\partial Q}{\partial t} = -\frac{\partial}{\partial \psi_1}[\Gamma(\psi_1-\psi_2)Q] - \frac{\partial}{\partial \psi_2}[\Gamma(\psi_2-\psi_1)Q] + D\left(\frac{\partial^2}{\partial \psi_1^2} + \frac{\partial^2}{\partial \psi_2^2}\right)Q, \tag{5.5.12}$$

where the definition of $\Gamma(\psi_\alpha - \psi_{\alpha'})$ is the same as in (5.2.17b), and $\bar{D}$ has been abbreviated as D. Unlike a coupled pair of oscillators without fluctuations which we treated in Sect. 5.2, case (b), the present model by no means shows perfect mutual entrainment. This is reminiscent of the fact that a pair of coupled magnetic spins with thermal fluctuation can show no phase transitions.

Let us look into what occurs in further detail. Let ψ_+ denote the mean phase and ψ_- the relative phase (i.e., $\psi_\pm = \psi_1 \pm \psi_2$) and $\tilde{Q}(\psi_-, t)$ the reduced distribution function with respect to the relative phase, or

$$\tilde{Q}(\psi_-, t) = \int_0^T Q(\psi_1, \psi_2, t)\,d\psi_+ .$$

By integrating (3.4.12) with respect to ψ_+, we have

$$\frac{\partial \tilde{Q}}{\partial t} = -\frac{\partial}{\partial \psi_-}\{[\Gamma(\psi_-) - \Gamma(-\psi_-)]\tilde{Q}\} + 2D\frac{\partial^2 \tilde{Q}}{\partial \psi_-^2}. \tag{5.5.13}$$

The implication of this equation is better understood by writing down the nonlinear Langevin equation equivalent to it:

$$\frac{\partial \psi_-}{\partial t} = -\frac{dV(\psi_-)}{d\psi_-} + f(t), \quad \text{where}$$

$$V(\psi_-) = -\int_0^{\psi_-}[\Gamma(x) - \Gamma(-x)]\,dx$$

and $f(t)$ is a Gaussian white noise function. The above looks like a dissipative stochastic motion of a particle in a periodic potential V. The particle will jump from one potential well to another, stochastically. If translated into the language of our oscillator system, this means that the oscillators are not capable of locking phase with each other.

The aforementioned magnetic-spin analogue strongly suggests that a phase-transition-like behavior may occur if we let the number of oscillators go to infinity. Thus the following case (Sect. 5.5.3) forms our central concern.

5.5.3 Many Oscillators Which are Statistically Identical

An appropriate system of equations would be

$$\frac{dX_\alpha}{dt} = F(X_\alpha) + \varepsilon\left[\sum_{\alpha' \neq \alpha} V_{\alpha\alpha'}(X_\alpha, X_{\alpha'}) + f_\alpha(X_\alpha, t)\right]. \tag{5.5.14}$$

We assume the same properties for the fluctuating forces as in the previous cases. Then

$$\frac{\partial Q}{\partial t} = -\sum_\alpha \frac{\partial}{\partial \psi_\alpha}\left[\sum_{\alpha'} \Gamma_{\alpha\alpha'}(\psi_\alpha - \psi_{\alpha'}) Q\right] + D\sum_\alpha \frac{\partial^2 Q}{\partial \psi_\alpha^2}. \tag{5.5.15}$$

How to treat this equation is the subject of the following two sections.

5.6 Statistical Model Showing Synchronization-Desynchronization Transitions

The phase-transition-like phenomena expected from system (5.5.15) seem to be far more difficult to treat than the usual thermodynamic phase transitions. This is simply because we do not know what is the steady solution, if any, of (5.5.15). In view of the fact that exact analytical results are seldom available even if the equilibrum distribution is known, we shall have to resort to some drastic assumptions about the interaction $\Gamma_{\alpha\alpha'}$ in order to obtain some specific results. In this connection, a Husimi-Temperly-type model

$$\Gamma_{\alpha\alpha'}(x) = N^{-1}\Gamma(x) \tag{5.6.1}$$

appears to be suitable again, although unlike the model used in Sect. 5.4, we need not assume here an explicit functional form of $\Gamma(x)$.

Let $n(\psi, t)$ represent the average number density, or

$$n(\psi, t) = \langle \hat{n}(\{\psi_\alpha\}; \psi)\rangle_t = \int_0^T \prod_\alpha d\psi_\alpha \cdot \hat{n}(\{\psi_\alpha\}; \psi)\, Q(\{\psi_\alpha\}, t), \tag{5.6.2}$$

where $\hat{n}$ is defined as in (5.4.6). Rather than treating the Fokker-Planck equation (5.5.15), it turns out more advantageous to work with the evolution equation for $n(\psi, t)$. For a general pair interaction $\Gamma_{\alpha\alpha'}(x)$, the equation for n would necessarily involve various many-body correlation functions, so that such a chain of equations would never be closed without invoking some truncation hypothesis.

In contrast, for interactions of infinitely long range like (5.6.1), it is possible to get a single equation for n involving no higher correlation functions. Note that the same idea does not, unfortunately, work for the deterministic system of Sect. 5.4. The derivation of the equation for n is made easier by making use of the identities

$$\frac{\partial}{\partial \psi_\alpha} \hat{n}(\{\psi_\alpha\}; \psi) = -N^{-1} \frac{\partial}{\partial \psi} \delta(\psi - \psi_\alpha) , \tag{5.6.3a}$$

$$\sum_{\alpha'} \Gamma(\psi_\alpha - \psi_{\alpha'}) = \sum_{\alpha'} \int_0^T d\psi' \Gamma(\psi_\alpha - \psi') \delta(\psi' - \psi_{\alpha'}) . \tag{5.6.3b}$$

Taking the time derivative of (5.6.2), and substituting (5.5.15) for $\partial Q/\partial t$, we have

$$\begin{aligned}\frac{\partial}{\partial t} n(\psi, t) = & -N^{-1} \int_0^T \prod_{\alpha''} d\phi_{\alpha''} \hat{n}(\{\psi_\alpha\}; \psi) \sum_\alpha \frac{\partial}{\partial \psi_\alpha} \left[\sum_{\alpha'} \Gamma(\psi_\alpha - \psi_{\alpha'}) Q \right] \\ & + D \int_0^T \prod_{\alpha'} d\psi_{\alpha'} n(\{\psi_\alpha\}; \psi) \sum_\alpha \frac{\partial^2 Q}{\partial \psi^2} .\end{aligned} \tag{5.6.4}$$

By making a partial integration, and using the properties (5.6.3a, b), this equation becomes

$$\begin{aligned}\frac{\partial}{\partial t} n(\psi, t) = & -N^{-1} \int_0^T \prod_{\alpha''} d\psi_{\alpha''} \sum_{\alpha, \alpha'} \left[N^{-1} \frac{\partial}{\partial \psi} \delta(\psi - \psi_\alpha) \right] \int_0^T d\psi' \Gamma(\psi_\alpha - \psi') \\ & \times \delta(\psi' - \psi_{\alpha'}) Q + N^{-1} D \sum_\alpha \int_0^T \prod_{\alpha'} d\psi_{\alpha'} \left[\frac{\partial^2}{\partial \psi^2} \delta(\psi - \psi_\alpha) \right] Q \\ = & -\frac{\partial}{\partial \psi} \int_0^T d\psi' \Gamma(\psi - \psi') \int_0^T \prod_{\alpha''} d\psi_{\alpha''} \left[N^{-1} \sum_\alpha \delta(\psi - \psi_\alpha) \right] \\ & \times \left[N^{-1} \sum_{\alpha'} \delta(\psi - \psi_{\alpha'}) \right] Q + D \frac{\partial^2}{\partial \psi^2} \sum_\alpha \int_0^T \prod_{\alpha''} d\psi_{\alpha''} \\ & \times \left[N^{-1} \sum_\alpha \delta(\psi - \psi_\alpha) Q \right] \\ = & -\frac{\partial}{\partial \psi} \int_0^T d\psi' \cdot \Gamma(\psi - \psi') \langle \hat{n}(\{\psi_\alpha\}; \psi) \hat{n}(\{\psi_\alpha\}; \psi') \rangle_t \\ & + D \frac{\partial^2}{\partial \psi^2} n(\psi, t) .\end{aligned} \tag{5.6.5}$$

Since $\hat{n}(\{\psi_\alpha\}; \psi)$ is a macrovariable, its fluctuation, defined by

$$\delta \hat{n} = \hat{n} - n ,$$

is expected to be of the order $1/\sqrt{N}$, and hence negligible. Thus, in (5.6.5), one is allowed to put

$$\langle \hat{n}(\{\psi_\alpha\};\psi)\hat{n}(\{\psi_\alpha\};\psi')\rangle_t = n(\psi,t)\,n(\psi',t)\,. \tag{5.6.6}$$

Consequently, we obtain a simple nonlinear diffusion equation,

$$\frac{\partial n}{\partial t} = -\frac{\partial}{\partial \psi}\int_0^T d\psi'\,\Gamma(\psi-\psi')\,n(\psi,t)\,n(\psi',t) + D\frac{\partial^2 n}{\partial \psi^2}\,. \tag{5.6.7}$$

This is subject to the periodic boundary condition

$$n(\psi+T,t) = n(\psi,t) \tag{5.6.8}$$

and the normalization condition

$$\int_0^T n(\psi,t)\,d\psi = 1\,. \tag{5.6.9}$$

Of course, (5.6.7) conserves $\int_0^T n(\psi,t)\,d\psi$ as it ought to. Recently, (5.6.7) aroused some mathematical interest, associated with its exact solutions for special classes of kernel Γ (Satsuma, 1981).

Two types of particular solutions of (5.6.7) are of interest to us, namely, the constant solution

$$n(\psi,t) = n_0 = T^{-1} \tag{5.6.10}$$

and the traveling solutions

$$n(\psi,t) = n(\psi-\Omega t)\,. \tag{5.6.11}$$

The next section treats the transition or the exchange of stability between them.

5.7 Bifurcation of Collective Oscillations

We let $\varrho(\psi,t)$ denote the density deviation, or

$$n(\psi,t) = n_0 + \varrho(\psi,t)\,.$$

From now on, the reference period T is adjusted to be 2π by adopting a suitable time scale. Thus, $n_0 = 1/2\pi$. Since $\varrho(\psi,t)$ and $\Gamma(\psi)$ are 2π-periodic functions of ψ, they may be expanded as

$$\varrho(\psi,t) = \frac{1}{2\pi}\sum_{l\neq 0}\varrho_l(t)\,\mathrm{e}^{\mathrm{i}l\psi}\,,$$

$$\Gamma(\psi) = \sum_l \Gamma_l e^{il\psi} .$$

The factor $1/2\pi$ in the first expression is only for later convenience. It is clear that $\varrho_l = \bar{\varrho}_{-l}$ and $\Gamma_l = \bar{\Gamma}_{-l}$, or

$$\Gamma_l' = \Gamma_{-l}', \qquad \Gamma_l'' = -\Gamma_{-l}'', \tag{5.7.1}$$

where Γ_l' and Γ_l'' are the real and imaginary parts of Γ_l, respectively. Equation (5.6.7) may then be expressed as a system of nonlinear mode-coupling equations

$$\frac{d\varrho_l}{dt} = \sigma_l \varrho_l - il \sum_{m \neq 0, l} \Gamma_m \varrho_m \varrho_{l-m}, \qquad l = \pm 1, \pm 2, \dots , \tag{5.7.2}$$

where

$$\sigma_l = (l\Gamma_l'' - l^2 D) - il(\Gamma_0 + \Gamma_l') .$$

The stability and bifurcation may be discussed in essentially the same way as in Chap. 2. Let D be taken as a control parameter, all the other parameters being kept constant. As we decrease D, the sign of $\mathrm{Re}\{\sigma_l\}$ changes from negative to positive, and this corresponds to the instability of the constant solution to the lth mode. Thus the critical value of D is given by

$$D_c = \lambda^{-1} \Gamma_\lambda'' , \tag{5.7.3}$$

where λ is a particular l corresponding to the largest $l^{-1}\Gamma_l''$. It is clear that a pair of modes with $l = \pm\lambda$ becomes unstable simultaneously because $\mathrm{Re}\{\sigma_l\} = \mathrm{Re}\{\sigma_{-l}\}$. Furthermore, we have, in general, $\mathrm{Im}\{\sigma_l\} = -\mathrm{Im}\{\sigma_{-l}\} \neq 0$, so that the bifurcation here is the Hopf type. Note that $D > 0$ by definition. Therefore, the trivial solution never loses stability if $l\Gamma_l'' < 0$ for every l. Let us inquire into the physical meaning of $l\Gamma_l'' < 0$. We suppose for simplicity that $\Gamma(\psi)$ contains only a single mode l. Then,

$$\left.\frac{d\Gamma}{d\psi}\right|_0 = il\Gamma_l - il\Gamma_{-l} = -2l\Gamma_l'' .$$

In the light of the discussion in Sect. 5.2.2, negative $l\Gamma_l''$ implies repulsive coupling. Thus the reason for the absence of a phase transition is clear.

Right at the threshold, D_c, the only undamped modes are the critical mode pair, so that, except for initial transients, the small amplitude $\varrho(\psi, t)$ is expected to contain only these components, or

$$\varrho(\psi, t) = \frac{1}{2\pi}(We^{i(\lambda\psi + \omega t)} + \text{c.c.}) , \tag{5.7.4}$$

where

$$\omega = -\lambda(\Gamma_0 + \Gamma_\lambda') .$$

If the amplitude W is a constant, then this expression for $\varrho(\psi, t)$ represents a traveling wave with velocity

$$\Omega = \Gamma_0 + \Gamma_\lambda' .$$

When D deviates slightly from D_c, one may still expect that the dominant part of $\varrho(\psi, t)$ preserves the form of (5.7.4) except that W should then be interpreted as a small-amplitude and slowly varying function of t. As shown below, the application of the theory of Sect. 2.2 reveals that W obeys the Stuart-Landau equation near criticality.

We define a bifurcation parameter μ through

$$D = D_c(1-\mu) ,$$

and define ε by $\varepsilon = \sqrt{|\mu|}$. Then

$$\sigma_l = \sigma_{l,0} \pm \varepsilon^2 \sigma_{l,2} , \quad \text{where} \tag{5.7.5}$$

$$\sigma_{l,0} = l\Gamma_l'' - l^2 D_c - \mathrm{i}l(\Gamma_0 + \Gamma_l') ,$$

$$\sigma_{l,2} = l^2 D_c .$$

We now introduce a scaled time τ, by $\tau = \varepsilon^2 t$. The time differentiation appearing in (5.7.2) should then be transformed as

$$\frac{d}{dt} \rightarrow \frac{\partial}{\partial t} + \varepsilon^2 \frac{\partial}{\partial \tau} . \tag{5.7.6}$$

We substitute (5.7.5) and the expansion

$$\varrho_l(t, \tau) = \varepsilon \varrho_{l,1}(t, \tau) + \varepsilon^2 \varrho_{l,2}(t, \tau) + \dots \tag{5.7.7}$$

into (5.7.2). After making the transformation (5.7.6), we obtain

$$\left(\frac{\partial}{\partial t} + \varepsilon^2 \frac{\partial}{\partial \tau}\right) \sum_{\nu=1}^{\infty} \varepsilon^\nu \varrho_{l,\nu} = (\sigma_{l,0} \pm \varepsilon^2 \sigma_{l,2}) \sum_{\nu=1}^{\infty} \varepsilon^\nu \varrho_{l,\nu}$$

$$- \mathrm{i}l \sum_{m \neq 0, l} \Gamma_m \sum_{\nu, \nu'=1}^{\infty} \varepsilon^{\nu+\nu'} \varrho_{m,\nu} \varrho_{l-m,\nu'} . \tag{5.7.8}$$

This equation is equivalent to the system of balance equations

$$\left(\frac{\partial}{\partial t} - \sigma_{l,0}\right) \varrho_{l,\nu} = B_{l,\nu} , \qquad \nu = 1, 2, \dots , \tag{5.7.9}$$

where

$$\sigma_{\pm\lambda,0} = \pm \mathrm{i}\omega , \qquad \mathrm{Re}\{\sigma_{l,0}\} < 0 \qquad (l \neq \pm\lambda) . \tag{5.7.10}$$

The inhomogeneous terms $B_{l,\nu}$ up to $\nu=3$ have the forms

$$B_{l,1}=0\,, \tag{5.7.11 a}$$

$$B_{l,2}=-\mathrm{i}l\sum_{m\neq 0,l}\Gamma_m\varrho_{m,l}\varrho_{l-m,l}\,, \tag{5.7.11 b}$$

$$B_{l,3}=-\left(\frac{\partial}{\partial\tau}\mp\sigma_{l,2}\right)\varrho_{l,1}-\mathrm{i}l\sum_{m\neq 0,l}\Gamma_m(\varrho_{m,2}\varrho_{l-m,1}+\varrho_{m,1}\varrho_{l-m,2})\,. \tag{5.7.11 c}$$

The solvability condition takes the form

$$\int_0^{2\pi/\omega}B_{\lambda,\nu}(t,\tau)\mathrm{e}^{-\mathrm{i}\omega t}dt=0\,. \tag{5.7.12}$$

If $B_{\lambda,\nu}$ are expanded into various harmonics as

$$B_{\lambda,\nu}(t,\tau)=\sum_{s=-\infty}^{\infty}B_{\lambda,\nu}^{(s)}(\tau)\mathrm{e}^{\mathrm{i}s\omega t}\,,$$

then (5.7.12) reduces to

$$B_{\lambda,\nu}^{(1)}=B_{-\lambda,\nu}^{(-1)}=0\,. \tag{5.7.13}$$

We now try to solve (5.7.9) iteratively. For $\nu=1$,

$$\varrho_{\lambda,1}=\bar{\varrho}_{-\lambda,1}=W(\tau)\,\mathrm{e}^{\mathrm{i}\omega t}\,, \tag{5.7.14a}$$

$$\varrho_{l,1}=0\qquad(l\neq\pm\lambda)\,, \tag{5.7.14b}$$

where $W(\tau)$ is the quantity yet to be determined. The inhomogeneous terms for $\nu=2$ then become

$$B_{l,2}=0\,,\qquad l\neq\pm 2\lambda\,, \tag{5.7.15a}$$

$$B_{2\lambda,2}=\bar{B}_{-2\lambda,2}=-2\mathrm{i}\lambda\Gamma_\lambda W^2\mathrm{e}^{2\mathrm{i}\omega t}\,. \tag{5.7.15b}$$

Thus the second-order deviations are found to be

$$\varrho_{2\lambda,2}=\bar{\varrho}_{-2\lambda,2}=-\frac{2\mathrm{i}\lambda\Gamma_\lambda}{2\mathrm{i}\omega-\sigma_{2\lambda,0}}W^2\mathrm{e}^{2\mathrm{i}\omega t}\,, \tag{5.7.16a}$$

$$\varrho_{l,2}=0\,,\qquad l\neq\pm 2\lambda,\pm\lambda\,; \tag{5.7.16b}$$

the $\varrho_{\pm\lambda,2}$ may be non-vanishing, but their explicit expressions are not necessary for our present purpose. As expected, the solvability condition (5.7.13) for $\nu=3$ takes the form of the Stuart-Landau equation,

$$B_{\lambda,3}^{(1)} = -\left(\frac{\partial}{\partial \tau} \mp \sigma_{\lambda,2}\right) W - g|W|^2 W = 0\,, \qquad \mu \gtrless 0\,, \tag{5.7.17}$$

where

$$g = \frac{2\lambda^2 \Gamma_\lambda(\Gamma_{-\lambda} + \Gamma_{2\lambda})}{2\mathrm{i}\omega - \sigma_{2\lambda,0}}\,. \tag{5.7.18}$$

The sign of g', i.e., the real part of g, depends on the coupling function $\Gamma(\psi)$. As a special case, suppose $\Gamma(\psi)$ contains a single pair of the Fourier components $\Gamma_{\pm 1}$, i.e.,

$$\Gamma(\psi) = -K \sin(\psi + \psi_0)\,,$$

$$K > 0\,, \quad |\psi_0| < \frac{\pi}{2}\,.$$

Then

$$g = \frac{\lambda |\Gamma_\lambda|^2}{2\Gamma_\lambda'' - \mathrm{i}\Gamma_\lambda'}\,, \tag{5.7.19}$$

which leads to $g' > 0$. In this particular case, the bifurcation is therefore super-critical, and the bifurcating solution is stable.

6. Chemical Waves

The phase description can explain expanding target patterns in reaction-diffusion systems. The same method, however, breaks down for rotating spiral waves because of a phase singularity involved. The Ginzburg-Landau equation is then invoked.

6.1 Synchronization in Distributed Systems

The assembly of coupled limit cycle oscillators easily shows organized motion. This seems to be especially true when the constituent oscillators are identical and their mutual coupling is of the attractive type. The system then behaves like a perfectly self-synchronized unit as we observed in Chap. 5.

Our previous consideration was, however, restricted to a special form of the coupling where the way in which the oscillators are distributed in physical space was entirely irrelevant. We now remove this restriction. Then, a natural question which may arise would be what variety one may expect for possible modes of ordered motion shown by the fields of identical oscillators coupled through *short*-range interactions, such as diffusion coupling of an attractive type. (Note that diffusion coupling may happen to be repulsive, as we saw in Sect. 5.3.) Due to the finiteness of the interaction range, local events could not be experienced by distant points without time lag. As a consequence, coherent dynamical modes in such distributed systems, if any, are expected to appear as various forms of synchronizing waves or chemical waves. The study of such waves is the subject of this chapter. For the repulsive-type diffusion coupling, some pathological dynamics may be expected. This leads, as we shall see in Chap. 7, to diffusion-induced chemical turbulence.

Most of the theoretical work on chemical waves has, in the past, been motivated by the experimental finding of wave patterns in the Belousov-Zhabotinsky reaction. For a general survey of this reaction system, see Zhabotinsky (1974) and Tyson (1976); the latter includes a readable presentation of the reaction mechanisms discoverded by Field, Körös, and Noyes (1972), and also how a three-variable model called the Oregonator is constructed. There are two representative types of wave patterns known in the Belousov-Zhabotinsky reaction, namely, expanding targetlike waves, and rotating spiral waves. It is a commonly accepted view that the appearance of such wave patterns is not due to some peculiarities of this reaction system but is feature common to reaction-diffusion systems of an oscillatory nature in general. Moreover, it was discovered by Winfree (1972) that a non-oscillatory version of the Belousov-Zhabotinsky reaction can also sustain the same wave patterns. Although not directly related to chemical reactions, the discovery of rotating waves in excitable media can be traced back to the experiment by Nagumo et al. (1962) (published only in Japanese; see also Suzuki, 1976) using a grid of iron wires in nitric acid, thereby simulating excitable nervelike tissues. In the field of physiology, there is a series of remarkable experiments by Allessie et al. (1977) who succeeded in producing rotating waves in rabbit atrial muscle, which is a typical excitable medium. As another interesting biological example, one may mention the cellular slime mold Dictyostelium discoideum, which can show both circular waves and spiral waves in a certain period of its life cycle (Gerisch, 1968).

In this chapter, chemical waves will be discussed by means of the nonlinear phase diffusion equation (3.3.5 or 6) and the Ginzburg-Landau equation (2.4.13). On account of some serious limitations involved in these model equations, we do not attempt to explain any experimentally observed patterns for the Belousov-Zhabotinsky reaction, but we only indicate some similarities existing between theoretically obtained wave patterns and real ones. In fact, it is very important to realize that under usual conditions, the waves in the Belousov-Zhabotinsky reaction are considerably different in nature from the waves obtained from our model systems. Realistic chemical waves are usually more like *trigger waves* which are characterized by their being blocked by impermeable barriers. It is known that such waves are characteristic to relaxation oscillations or, more typically, non-oscillating excitable kinetics. In contrast, the waves described by the nonlinear phase diffusion equation are more like so-called *phase waves* which have a slow space dependence and are hardly blocked by impermeable barriers (Kopell and Howard, 1973b; Winfree, 1974b). The waves obtained from the Ginzburg-Landau equation are also different in nature from trigger waves because the local oscillators described by this equation show completely smooth oscillations. It is rather surprising, therefore, that the wave patterns to be obtained theoretically below share many important features in common with the real ones. A merit of our theoretical approach would be its mathematical simplicity, which enables us to treat different patterns in a unified manner.

A number of research works have so far appeared concerning more sophisticated treatments of specific wave phenomena. To mention a few of them, Ortoleva and Ross (1973) gave an explanation of target pattern in terms of phase waves; Howard and Kopell (1977) discussed the formation of shocklike structures resulting from wave collisions; Greenberg (1976) made a study of the general asymptotic structure of rotating-spiral and axisymmetric solutions far from the central core; Cohen, Neu and Rosales (1978) developed an analytical theory to obtain spiral-wave solutions, including the core region, by means of the $\lambda-\omega$ system. There is also the work by Greenberg (1980) in the same direction; Erneux and Herschkowitz-Kaufman (1977) employed a bifurcation-theoretical approach combined with numerical calculations, to obtain rotating waves; Fife (1979b) developed a theory based on a singular perturbation method which is suited to trigger waves, and discussed the formation of targetlike patterns. In addition to these analytical works, we mention Winfree's numerical simulation and qualitative arguments on rotating waves, which used a simple excitable kinetics, and which were quite instrumental in clarifying the structure of the central core as a phase singularity (Winfree, 1978). The theory of this chapter originates from the work by Kuramoto and Yamada (1976b), and Yamada and Kuramoto (1976a), and is partly in common with the work of Ortoleva and Ross (1973), Howard and Kopell (1977), Greenberg (1976), and Cohen et al. (1978).

6.2 Some Properties of the Nonlinear Phase Diffusion Equation

As a preliminary to investigating two-dimensional wave patterns, we first make a brief inspection of some particular solutions of the nonlinear phase diffusion equation in one dimension,

$$\frac{\partial \psi}{\partial t} = \alpha \frac{\partial^2 \psi}{\partial x^2} + \beta \left(\frac{\partial \psi}{\partial x} \right)^2 , \tag{6.2.1}$$

and clarify their physical implications. A positive sign is assumed for α. As to the sign of β, see below. It is easy to confirm that (6.2.1) has a family of solutions with linear space-time dependence:

$$\psi(x,t) = \beta q^2 t + qx + \psi_0 , \tag{6.2.2}$$

where q and ψ_0 are arbitrary constants. Note, however, that large q values violate the very assumption on which the derivation of (6.2.1) itself is based. Remembering that the composition vector X was previously approximated as $X(x,t) \simeq X_0(t+\psi)$, where X_0 is a T-periodic function of $t+\psi$, we see that the solution (6.2.2) yields a family of periodic waves of X parametrized by q. Although the true wavenumber is $T^{-1}q$, we will still call q the wavenumber for simplicity. These waves are clearly stable, as far as $\alpha > 0$. Empirically, the appearance of periodic waves makes local frequencies higher than the frequency of the uniform oscillation, which implies that β should be taken to be positive. This property cannot generally be derived theoretically, and we simply assume it below.

There exists a slightly more general class of solutions of (6.2.1) than (6.2.2), and they are obtained by smoothly joining a pair of periodic waves of different q. To find such solutions, we first note that (6.2.1) is equivalent to the Burgers equation (Burgers, 1974)

$$\frac{\partial v}{\partial t} = \alpha \frac{\partial^2 v}{\partial x^2} - v \frac{\partial v}{\partial x} , \tag{6.2.3}$$

via the transformation

$$v = -2\beta \frac{\partial \psi}{\partial x} . \tag{6.2.4}$$

It is also a well-known fact that the Burgers equation (6.2.3) has a family of shock solutions

$$v(x,t) = v_0 + v_1 \tanh\left[-\frac{v_1}{2\alpha}(x - v_0 t)\right], \tag{6.2.5}$$

where v_0 and v_1 are arbitrary constants. The corresponding ψ is given by

$$\psi(x,t) = \frac{\alpha}{\beta}\ln\left\{\cosh\left[\frac{b\beta}{\alpha}(x + 2a\beta t)\right]\right\} + ax + \beta(a^2 + b^2)t, \tag{6.2.6}$$

where the last term linear in t (i.e., the constant of integration of $\int v\,dx$) was so determined that ψ satisfies (6.2.1); a and b are related to v_0 and v_1 through $v_0 = -2a\beta$ and $v_1 = -2b\beta$. The position x_s of the shock front is given by

$$x_s = -2a\beta t. \tag{6.2.7}$$

To see some physical implications of (6.2.6), it would be better to work with the original phase variable $\phi(= t + \psi)$ rather than ψ. Then (6.2.6) says that ϕ behaves like

$$\phi \simeq \phi_\pm \equiv \Omega_\pm t + q_\pm x \tag{6.2.8}$$

far from the shock (i.e., $x - x_s \to \pm\infty$), where

$$q_\pm = a \pm |b|, \qquad \Omega_\pm = 1 + \beta q_\pm^2.$$

It is now clear that (6.2.6) represents a two-parameter family of solutions, each representing a junction of two periodic waves. It is easy to confirm that

$$\Omega_+ \gtrless \Omega_-, \qquad \frac{dx_s}{dt} \lessgtr 0, \qquad \text{according to } a \gtrless 0.$$

Thus the shock front moves in such a way that the higher-frequency domain expands at the expense of the lower-frequency domain; eventually, the entire system will be dominated by the former. This feature is characteristic of the entrainment phenomena for spatially distributed oscillators. The formation of target patterns is considered to be a two-dimensional version of such phenomena (Sect. 6.3). The propagation velocities of the periodic waves in the right and left regions are given by

$$c_\pm = -\Omega_\pm / q_\pm. \tag{6.2.9}$$

Suppose that the directions of propagation are opposite in these regions, i.e., $c_+ c_- < 0$ and hence $q_+ q_- < 0$. Then the shock front is seen to represent a wave sink (i.e., $c_+ < 0$, $c_- > 0$) and not a source (i.e., $c_+ > 0$, $c_- < 0$).

6.3 Development of a Single Target Pattern

Expanding target patterns in the Belousov-Zhabotinsky reaction were first discovered by Zaikin and Zhabotinsky (1970). In their experiment, they used a spontaneously oscillating thin layer of solution which was contained in a Petri dish of diameter 100 mm. The reagent contained bromate, bromomalonic acid and ferroin. With this prescription, one may observe periodic alternation of oxidized and reduced forms of the catalyst through a dramatic color change of the solution between red (reduced state) and blue (oxidized state). Some features of the pattern observed by them and by later experimenters are the following:

1) Blue waves are sent out periodically from an isolated point (called a pacemaker), and they propagate outward in the form of concentric rings or a target pattern.
2) Some such target patterns may coexist in the same medium. The frequencies at which the waves are emanated, which we call the pacemaker frequencies, differ from pacemaker to pacemaker, but they are definitely higher than the frequency of the uniform oscillation.
3) The propagation velocity of the waves is approximately the same for all target patterns which coexist. These waves can be blocked by impermeable barriers, implying that they represent trigger waves.
4) The waves annihilate each other on head-on collisions. Consequently, a pair of colliding circular waves forms an angular structure in the wave front.
5) The outermost ring of a given target pattern disappears once per background-oscillation period. Nevertheless, the pattern itself continues to expand since the rate at which new waves are sent out from the pacemaker center is higher than the rate of wave annihilation.
6) In the course of successive collisions of the waves, the domain of the faster pacemaker invades step by step the domain of the slower pacemaker. Eventually, we are left with a single target pattern which has the fastest pacemaker of all. Note, however, that this feature is a direct consequence of the constancy of the propagation velocity.

Experimentally, the pacemakers seem to arise from heterogeneous nuclei (e.g., dust particles, scratches on the vessel, etc.), in the vicinity of which the oscillation frequency seems to be made somewhat higher than that of the bulk oscillation. In fact, if one carefully eliminates heterogeneities, e.g., by filtering the solution, then no target patterns can appear. In order to discuss target patterns mathematically, it therefore seems appropriate to take account of heterogeneities explicitly in the evolution equations. This was first done by Ortoleva and Ross (1973) by treating them as a perturbation.

Let us now apply the method of phase description I to the present problem by assuming that both the heterogeneities and diffusion are perturbations of order ε. Equation (3.3.1) should then be generalized as

$$\varepsilon p(X) = D \nabla^2 X + G(X, r) , \tag{6.3.1}$$

where G represents the effect of some heterogeneities. In the lowest-order approximation, one may immediately get (after putting $\varepsilon = 1$) a modified form of the nonlinear phase diffusion equation,

$$\frac{\partial \phi}{\partial t} = 1 + \alpha \nabla^2 \phi + \beta (\nabla \phi)^2 + g(r), \quad \text{where} \tag{6.3.2}$$

$$g(r) = \frac{1}{T} \int_0^T d\phi Z(\phi) G(X_0(\phi), r).$$

Remember that α and β were assumed to be positive. It is appropriate to suppose that $g(r)$ is non-vanishing only in the vicinity of some points $r_1, r_2, \ldots$. One is now ready to discuss target patterns on the basis of (6.3.2).

Consider the nonlinear transformation

$$\phi(r,t) = \beta^{-1} \alpha \ln [Q(r,t)]. \tag{6.3.3}$$

This is essentially the same as the well-known Hopf-Cole transformation which reduces the Burgers equation to a simple diffusion equation (Burgers, 1974). The same transformation reduces (6.3.2) to the linear equation

$$\frac{\partial Q}{\partial t} = \alpha [\beta \alpha^{-2} + \nabla^2 - U(r)] Q, \quad \text{where} \tag{6.3.4}$$

$$U(r) = -\beta \alpha^{-2} g(r).$$

By definition, Q must be positive over the entire space-time domain. Note that multiplication of Q by a constant merely produces a trivial shift of ϕ.

Let a fundamental solution of (6.3.4) be expressed as

$$Q(r,t) = \hat{Q}(r) \exp [(\beta \alpha^{-1} + \alpha \lambda) t]. \tag{6.3.5}$$

This leads to the eigenvalue problem

$$\lambda \hat{Q}(r) = [\nabla^2 - U(r)] \hat{Q}(r). \tag{6.3.6}$$

The general solution of (6.3.4) may be expressed as

$$Q(r,t) = \sum_n \hat{Q}_n(r) \exp [(\beta \alpha^{-1} + \alpha \lambda_n) t], \tag{6.3.7}$$

where $\hat{Q}_n(r)$ is subject to the restriction

$$\sum_n \hat{Q}_n(r) = Q(r,0) > 0. \tag{6.3.8}$$

[The above inequality ensures $Q(r,t) > 0$ for all t.] It should be noted that the particular solution (6.3.5) with vanishing λ represents a steady solution. As far as this solution remains stable, the effects of heterogeneities would be of little interest since they could only give rise to some localized, finite-amplitude deformation in the phase profile. The target pattern seems to arise from the instability of such a phase-deformed steady state or, equivalently, from the appearance of a positive eigenvalue λ in (6.3.6). An interesting feature of the present problem is, unlike ordinary instability and bifurcation problems, that the linear instability need not be saturated by nonlinear effects. In fact, the indefinite exponential growth of Q is not an unphysical feature since it simply corresponds to the indefinite expansion of a target pattern. The eigenvalue λ is always real; note that (6.3.6) is identical to a quantum-mechanical potential problem with energy eigenvalues

$$E = -\lambda \tag{6.3.9}$$

in a system of units where $2m = 1$ and $\hbar = 1$. In quantum-mechanical terminology, a target pattern arises if the potential U admits a bound state. This can only happen for attractive potentials, or for heterogeneities that make the local frequency higher than that of the bulk oscillation. This feature of our model is consistent with property (2) of real target patterns.

In this section, our consideration is restricted to the effect of a single heterogeneous nucleus which produces a spherically symmetric potential $U(\boldsymbol{r}) = U(r)$, where U is assumed to be non-vanishing only near $r = 0$. The spatial dimension is assumed to be *three*. Although the aforementioned features (1 – 4) concern two-dimensional wave patterns, one may interpret them as the features of a two-dimensional intersection of three-dimensional patterns, where the pacemakers are supposed to lie almost on a common plane of intersection. Rather than considering a genuine two-dimensional system, we want to take this view partly because of the mathematical simplicity of working with exponential or trigonometric functions rather than with Bessel functions; the other reason is that such a view seems to be more realistic in that it takes account of the diffusion in the vertical direction, whose effect should not be neglected even if the chemical solution forms a thin layer. In accordance with the observed wave patterns, we may also restrict consideration to spherically symmetric solutions of (6.3.4). Targetlike patterns with lower symmetry have never been observed, which seems to be due to the fact that the ground-state (i.e., the s state) contribution is dominant in the summation in (6.3.7) as will be explained later. A certain angle-dependent solution (for the case of vanishing U) becomes important in Sect. 6.5 in connection with rotating waves.

Equation (6.3.6) is now reduced to

$$\lambda \hat{Q}(r) = [\nabla_r^2 - U(r)]\,\hat{Q}(r)\,, \quad \text{where} \tag{6.3.10}$$

$$\nabla_r^2 \equiv \frac{d^2}{dr^2} + \frac{2}{r}\,\frac{d}{dr}\,.$$

It is reasonable to require $\hat{Q}(r)$ to satisfy the following conditions:

$$\hat{Q}(r) > 0 \quad \text{everywhere}, \tag{C.1}$$

$$\hat{Q}(r) \quad \text{is finite as} \quad r \to 0, \infty. \tag{C.2}$$

The eigenstates which do not satisfy (C.1) are uninteresting because the corresponding terms in the summation in (6.3.7) could never be dominant due to the constraint that $Q(r,t) > 0$ everywhere. Without condition (C.2), $Q(r,t)$ would not be bounded even for finite t, which in turn implies that $Q(r,0)$ would also be unbounded. But what we actually want to know is the evolution process of a pattern starting from a uniformly oscillating state for which $Q(r,0)$ is obviously bounded.

The transformation

$$\hat{Q}(r) = \chi(r)/r \tag{6.3.11}$$

reduces (6.3.10) to a one-dimensional problem:

$$E\chi = \left[-\frac{d^2}{dr^2} + U(r) \right] \chi, \tag{6.3.12}$$

where E has been defined by (6.3.9). Although it would be difficult to know what would be a realistic functional form of U, this does not cause problems, since what is important is only the asymptotic form of the wave functions χ far from the pacemaker region. Thus, one may assume some simple form for $U(r)$ so that (6.3.12) may be explicitly solved. A most convenient choice would be a square-well potential. Let d denote its depth and σ the radius. Then,

$$\begin{aligned} E\chi &= -\frac{d^2\chi}{dr^2}, \quad r > \sigma, \\ (E+d)\chi &= -\frac{d^2\chi}{dr^2}, \quad r < \sigma. \end{aligned} \tag{6.3.13}$$

Let the sign of d be unspecified at first; the potential may be either attractive or repulsive. Note that the solutions with positive E are excluded because of (C.1). Furthermore, no solutions can simultaneously satisfy the inequalities $E+d<0$ and $E<0$, because such solutions would violate (C.2). We are then left with the following three possibilities:

(A) $-d < E < 0$

(B) $d > 0, \quad E = 0$

(C) $d < 0, \quad E = 0.$

Note also that the only possible solution is the zero-eigenvalue solution for $d < 0$. In more physical terms, the heterogeneous nuclei with a frequency lower than that of the bulk oscillation are not capable for forming a target pattern. Some elementary calculations give the following results, where we have used the notations χ_+ and χ_0 to indicate the eigenfunctions with positive λ (negative E) and vanishing λ, respectively:

$$\text{(A)} \qquad \chi_+ = \mathrm{e}^{-\sqrt{|E|}r}, \qquad r > \sigma, \tag{6.3.14a}$$

$$= a\sin(\sqrt{E+d}r), \qquad r < \sigma, \tag{6.3.14b}$$

$$a = \mathrm{e}^{-\sqrt{|E|}\sigma}/\sin(\sqrt{E+d}\sigma), \tag{6.3.14c}$$

$$-\sqrt{|E|} = \sqrt{E+d}\cot(\sqrt{E+d}\sigma). \tag{6.3.14d}$$

$$\text{(B)} \quad \chi_0 = r + b, \qquad r > \sigma, \tag{6.3.15a}$$

$$= b'\sin(\sqrt{d}r), \qquad r < \sigma, \tag{6.3.15b}$$

$$b = \sigma\left(\frac{\tan(\sqrt{d}\sigma)}{\sqrt{d}\sigma} - 1\right), \tag{6.3.15c}$$

$$b' = (1/\sqrt{d})\cos(\sqrt{d}\sigma). \tag{6.3.15d}$$

$$\text{(C)} \quad \chi_0 = r + c, \qquad r > \sigma, \tag{6.3.16a}$$

$$= c'\sinh(\sqrt{|d|}r), \qquad r < \sigma, \tag{6.3.16b}$$

$$c = \frac{\tanh(\sqrt{|d|}\sigma)}{\sqrt{|d|}\sigma} - 1, \tag{6.3.16c}$$

$$c' = (1/\sqrt{|d|})\cosh(\sqrt{|d|}\sigma). \tag{6.3.16d}$$

In each case one may arbitrarily multiply $\chi_{+,0}$ by a constant, which only adds a trivial constant to ϕ.

When the potential is attractive ($d > 0$), one may know from the eigenvalue equation (6.3.14d) the condition for the existence of bound states, i.e., the condition for the instability of the stationary Q. A too weakly attractive potential is unable to support bound states. This implies that the heterogeneities with strength below some critical value are unable to entrain the surrounding medium to a higher-frequency state; conversely, such heterogeneities would be entrained by the surroundings to the frequency of the bulk oscillation. The existence of a critical condition for the appearance of a target pattern is a most remarkable feature of the present theory. On putting

$$\sqrt{E+d}\sigma = \xi,$$

(6.3.14d) becomes

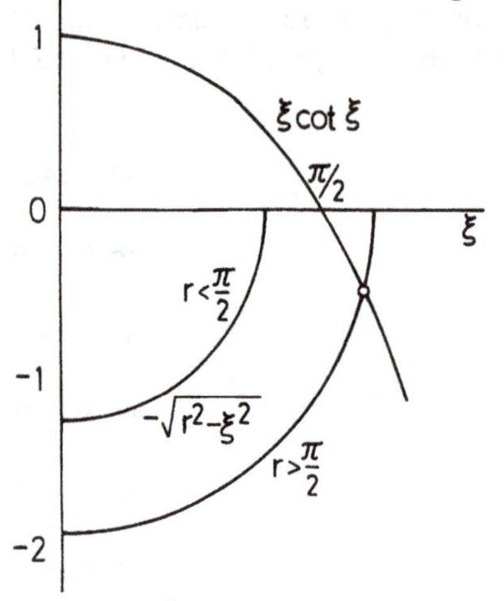

Fig. 6.1. Graphical solution of (6.3.17)

$$-\sqrt{d\sigma^2-\xi^2}=\xi\cot\xi\,. \tag{6.3.17}$$

One may find graphically the critical condition (Fig. 6.1); it is given by

$$\sqrt{d}\sigma=\frac{\pi}{2}\,. \tag{6.3.18}$$

The stationary value of Q is stable if $\sqrt{d}\sigma<\pi/2$ and unstable otherwise.

We now consider what occurs after instability. Let $Q(r,t)$ be expressed as a superposition of the zero-eigenvalue state and all bound states:

$$Q(r,t)=a_0\exp(\beta\alpha^{-1}t)\,\hat{Q}_0(r)+\sum_n a_n\exp[(\beta\alpha^{-1}+\alpha\lambda_n)t]\,\hat{Q}_n(r)\,, \tag{6.3.19}$$

where the a_n are positive and the suffix n specifies bound states. Corresponding to the outer solutions in (6.3.14a, 15a), we have

$$\hat{Q}_0=1+br^{-1}\quad\text{and} \tag{6.3.20}$$

$$\hat{Q}_n=r^{-1}\exp(-\sqrt{\lambda_n}r)=\exp(-\sqrt{\lambda_n}r-\ln r)\,. \tag{6.3.21}$$

Of particular interest is the asymptotic behavior of Q far from the pacemaker center (i.e., $r\gg\sigma$). Then the terms r^{-1} and $\ln r$ may be neglected in comparison with 1 and r, respectively. Thus,

$$Q(r,t)\simeq a_0\exp(\beta\alpha^{-1}t)\left\{1+\sum_n\exp[\alpha\lambda_n(t-t_n)-\sqrt{\lambda_n}r]\right\}, \tag{6.3.22}$$

where

$$t_n=-\ln(a_n/a_0)/\alpha\lambda_n\,.$$

Substituting (6.3.22) into (6.3.3), and dropping the trivial phase constant, we have

$$\phi(r,t) = t+\beta^{-1}\alpha\ln\left\{1+\sum_n \exp[\alpha\lambda_n(t-t_n)-\sqrt{\lambda_n}r]\right\}. \tag{6.3.23}$$

Suppose first that there is only one bound state. Then,

$$\phi(r,t) = t+\beta^{-1}\alpha\ln[1+\exp(\alpha\lambda t-\sqrt{\lambda}r)]. \tag{6.3.24}$$

The above implies the existence of a characteristic radius $R(t)$ defined by

$$\alpha\lambda t = \sqrt{\lambda}R(t). \tag{6.3.25}$$

A non-trivial asymptotic form of ϕ may be obtained by making both t and r tend to infinity, keeping r/t finite. If the wave pattern is seen coarsely on such a long spatial scale, the inversion of the numerical order between the two terms in the logarithm in (6.3.24) would be seen to occur suddenly, so that one may approximate it as

$$\phi(r,t) = t+\beta^{-1}\alpha\,\mathrm{Max}(0,\alpha\lambda t-\sqrt{\lambda}r), \quad \text{or} \tag{6.3.26}$$

$$\begin{aligned}\phi &= (1+\beta q^2)t-qr, & r<R(t),\\ &= t, & r>R(t),\end{aligned} \tag{6.3.27}$$

where $q=\beta^{-1}\alpha\sqrt{\lambda}$. It would be interesting to notice the analogy between (6.3.27) and (6.2.8). The oscillation frequency in the region $r<R(t)$ is higher than that in the region $r>R(t)$, which means that the former region has already been entrained to the pacemaker in order to acquire a frequency uniformly higher than the natural frequency. The entrained region is spherical and is expanding at a constant rate of radial increase. Inside the entrained sphere, waves propagate outward in concentric spheres whose velocity c is given by

$$c = q^{-1}(1+\beta q^2). \tag{6.3.28}$$

The property $c\to\infty$ as $q\to 0$ is characteristic of phase waves, and contrasts with the experimentally observed feature (3).

A convenient geometrical method for constructing a developing pattern from (6.3.26) is now described. Below, our interest will be in a two-dimensional intersection of the three-dimensional pattern, where the pacemaker is supposed to lie on the plane P of intersection. Let $\tilde{\phi}$ be defined by

$$\tilde{\phi} = \phi - t. \tag{6.3.29}$$

According to (6.3.26), the graph of $\tilde{\phi}$ on P looks like Fig. 6.2; for $r>R(t)$ it coincides with the plane P and stays still, while for $r<R(t)$ it is represented by the surface of a growing cone. It is convenient to imagine this as if P is inter-

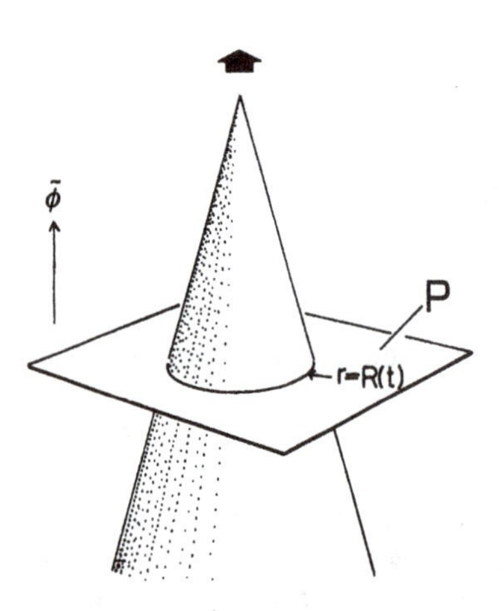

Fig. 6.2. Conical distribution of phase $\tilde{\phi}$. $\tilde{\phi}$ is growing in the region $r<R(t)$ and remains constant for $r>R(t)$. However, one may conveniently imagine the cone as penetrating the plane P and going upward at a constant speed

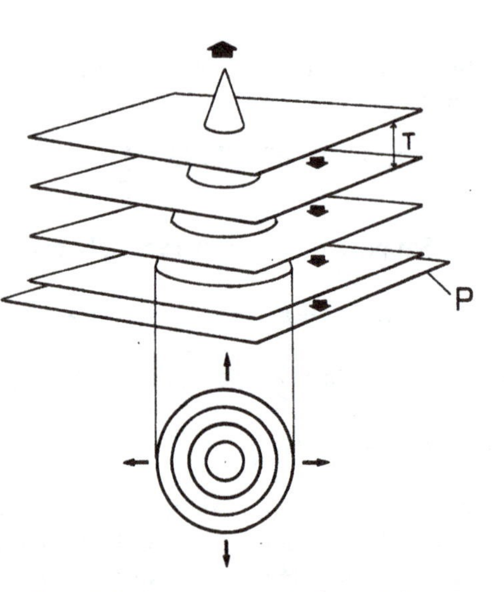

Fig. 6.3. Uniformly spaced horizontal planes intersecting a growing cone. The projection of the intersections onto the plane P forms an expanding target pattern

sected by a bottomless cone which is moving upward at a constant velocity βq^2. Let (r, θ) be the polar coordinates on P, the pacemaker being situated at $r=0$. The contours $r=r(\theta, t)$ of constant phase are then obtained from

$$\tilde{\phi}(r, \theta, t) = -t + nT. \tag{6.3.30}$$

A possible geometrical interpretation of this expression is the following: imagine a number of horizontal planes which are uniformly spaced by a distance T. These planes are descending at unit velocity. In general, we have some intersections of these planes with the graph of $\tilde{\phi}$. These intersections, if projected onto P, form equi-concentration or equi-phase contours (Fig. 6.3). Obviously, they form expanding concentric rings. It is seen that a ring is generated from the pacemaker center each time a descending plane comes to touch the top of the cone. On the other hand, a ring disappears when its radius attains the value $R(t)$; this occurs each time a descending plane arrives at the bottom of the cone P, i.e., once in a period T of the background oscillation. The same feature is also observed in real target patterns.

Finally, a brief comment on the case of multiple bound states is made. Suppose there are two bound states whose eigenvalues are λ_1 and $\lambda_2\,(<\lambda_1)$. Quite analogously to (6.3.26), we have the following approximate expression for ϕ:

$$\phi = t + \beta^{-1}\alpha\,\mathrm{Max}\,[0, \alpha\lambda_1(t-t_1) - \sqrt{\lambda_1}r, \alpha\lambda_2(t-t_2) - \sqrt{\lambda_2}r]\,. \tag{6.3.31}$$

There are three values of r for which a partial inversion of the numerical order occurs among the terms in the square bracket. They are

$$R_1(t) = \alpha\sqrt{\lambda_1}(t-t_1)\,,$$
$$R_2(t) = \alpha\sqrt{\lambda_2}(t-t_2)\,,$$
$$R_{12}(t) = \alpha\frac{\lambda_1(t-t_1)-\lambda_2(t-t_2)}{\sqrt{\lambda_1}-\sqrt{\lambda_2}}\,.$$

It is easy to see that the relation $R_{12} > R_1 > R_2$ holds for sufficiently large t. This immediately leads to the following property: if $r > R_{12}$ [i.e., if $\alpha\lambda_2(t-t_2) - \sqrt{\lambda_2}r > \alpha\lambda_1(t-t_1) - \sqrt{\lambda_1}r$], then $r > R_2$ [i.e., $\lambda_2(t-t_2) - \sqrt{\lambda_2}r < 0$], which says that the last term in the bracket in (6.3.31) can never be the largest. Essentially the same reasoning applies to the cases of multiple bound states in general, and one is led to the formula

$$\phi = t + \beta^{-1}\alpha\,\mathrm{Max}(0, \alpha\lambda_{max}t - \sqrt{\lambda_{max}}r)\,, \qquad (6.3.32)$$

where λ_{max} denotes the largest eigenvalue.

6.4 Development of Multiple Target Patterns

The argument above may be readily extended to include the situation where a number of target patterns coexist. We restrict consideration to a two-pacemaker problem since this demonstrates all essential points of theoretical interest. Let the potential $U(r)$ be of the form

$$U(\boldsymbol{r}) = U_A(|\boldsymbol{r}-\boldsymbol{r}_A|) + U_B(|\boldsymbol{r}-\boldsymbol{r}_B|)\,, \qquad (6.4.1)$$

where U_A and U_B are square-well potentials each characterized by the depth parameter d_α and radius $\sigma_\alpha(\alpha = A, B)$. Since only the pacemakers capable of forming patterns are of interest at present, d_A and d_B are assumed positive and large enough to have bound states. For simplicity, let U_α have a single bound state. This restriction could be removed in the same manner as in the last paragraph of the previous section. We already know the eigenfunctions for the system of a single square-well potential U_α. They are

$$\hat{Q}_0^\alpha = 1 + b|\boldsymbol{r}-\boldsymbol{r}_\alpha|^{-1}\,, \qquad |\boldsymbol{r}-\boldsymbol{r}_\alpha| > \sigma_\alpha\,, \qquad (6.4.2\text{a})$$

$$= b'\sin(\sqrt{d_\alpha}|\boldsymbol{r}-\boldsymbol{r}_\alpha|)/|\boldsymbol{r}-\boldsymbol{r}_\alpha|\,, \qquad |\boldsymbol{r}-\boldsymbol{r}_\alpha| < \sigma_\alpha\,, \qquad (6.4.2\text{b})$$

$$\hat{Q}_+^\alpha = \exp(-\sqrt{\lambda_\alpha}|\boldsymbol{r}-\boldsymbol{r}_\alpha|)/|\boldsymbol{r}-\boldsymbol{r}_\alpha|\,, \qquad |\boldsymbol{r}-\boldsymbol{r}_\alpha| > \sigma_\alpha\,, \qquad (6.4.3\text{a})$$

$$= a\sin(\sqrt{d_\alpha - \lambda_\alpha}|\boldsymbol{r}-\boldsymbol{r}_\alpha|)/|\boldsymbol{r}-\boldsymbol{r}_\alpha|\,, \qquad |\boldsymbol{r}-\boldsymbol{r}_\alpha| < \sigma_\alpha\,, \qquad (6.4.3\text{b})$$

where a, b, and b' are defined as before, see (6.3.14c, 15c, d). The above expressions are useful for considering the systems of two square-well potentials. For simplicity, let the pacemakers be sufficiently far apart from each other. Then the bound state Q_+ of the two-pacemaker problem may be approximated by a superposition of Q_+^A and Q_+^B. This is because $\hat{Q}_+^\alpha$ has a negligible value near $r_\beta(\neq\alpha)$ due to the exponential decay of $\hat{Q}_+^\alpha$. Thus,

$$Q_+(r,t) = c_A\hat{Q}_+^A(|r-r_A|)\exp[(\beta\alpha^{-1}+\alpha\lambda_A)t] \\ +c_B\hat{Q}_+^B(|r-r_B|)\exp[(\beta\alpha^{-1}+\alpha\lambda_B)t]\,, \tag{6.4.4}$$

where c_A and c_B are arbitrary positive constants. In contrast to Q_+, the zero-eigenvalue state Q_0 cannot be expressed as a simple superposition like this because $\hat{Q}_0^\alpha$ includes a constant term. Nevertheless, we can easily confirm that a slightly modified expression, as follows, approximately satisfies (6.3.4):

$$Q_0(r,t) = [1+b(|r-r_A|^{-1}+|r-r_B|^{-1})]\,e^{\beta\alpha^{-1}t}\,, \quad |r-r_A|>\sigma_A \quad \text{and} \quad |r-r_B|>\sigma_B\,, \tag{6.4.5a}$$

$$= b'[\sin(\sqrt{d_A}|r-r_A|)/|r-r_A|]\,e^{\beta\alpha^{-1}t}\,, \quad |r-r_A|<\sigma_A\,, \tag{6.4.5b}$$

$$= b'[\sin(\sqrt{d_B}|r-r_B|)/|r-r_B|]\,e^{\beta\alpha^{-1}t}\,, \quad |r-r_B|<\sigma_B\,. \tag{6.4.5c}$$

Most of such complexities in constructing correct two-pacemaker solutions are swept away by restricting attention to the asymptotic behavior of Q as $|r-r_\alpha|\to\infty$ and $t\to\infty$ with $|r-r_\alpha|/t$ finite. In fact, Q is then simplified as

$$Q(r,t) = a_+Q_+(r,t)+a_0Q_0 \\ \simeq a_0 e^{\beta\alpha^{-1}t}\{1+\exp[\alpha\lambda_A(t-t_A)-\sqrt{\lambda_A}|r-r_A|] \\ +\exp[\alpha\lambda_B(t-t_B)-\sqrt{\lambda_B}|r-r_B|]\}\,, \quad \text{or} \tag{6.4.6}$$

$$\phi(r,t) = t+\beta^{-1}\alpha\,\mathrm{Max}\,[0,\alpha\lambda_A(t-t_A)-\sqrt{\lambda_A}|r-r_A|,\alpha\lambda_B(t-t_B) \\ -\sqrt{\lambda_B}|r-r_B|]\,. \tag{6.4.7}$$

These expressions are quite analogous to (6.3.22) and (6.3.26), respectively. The quantities t_A and t_B may be interpreted as the onset times of the activity of the pacemakers. In order to discuss a wave pattern from the expression for ϕ above, the geometrical method described in the previous section is helpful. Let (x,y) be the coordinates of P (i.e., the plane of intersection of the three-dimensional wave pattern); both the pacemakers A and B are assumed to lie on P. Consider the graph of $\tilde{\phi}(x,y,t)$. Figure 6.4a shows a typical case where two target patterns are developing independently of each other. Eventually, the corresponding cones will come into contact with each other, and then they will begin to merge (Fig. 6.4b). One may construct contours of constant phase by intersecting the merged cones

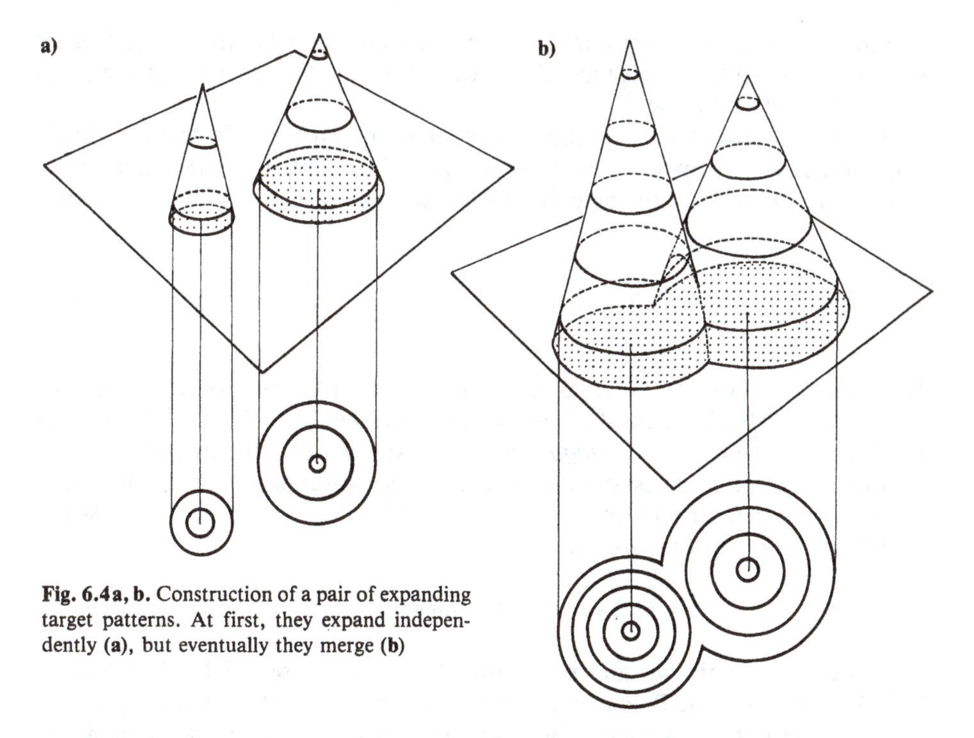

Fig. 6.4a, b. Construction of a pair of expanding target patterns. At first, they expand independently (**a**), but eventually they merge (**b**)

with uniformly spaced horizontal planes. It is clearly seen that the waves cancel each other by collisions, and some angular structures in the wave front are thereby formed. In our figure the frequency has been assumed to be higher for the left pacemaker. This means that the left cone (imagined again as bottomless and penetrating through up P) goes upward at velocity greater than that of the right cone. As a result, the right cone will eventually be swallowed by the left cone, leaving a single pattern in the system. The same features are also known for real target patterns.

6.5 Phase Singularity and Breakdown of the Phase Description

Rotating spiral waves in the Belousov-Zhabotinsky reaction were nicely analyzed and interpreted by Winfree (1972, 1974a). (For a general survey, see Winfree, 1978 and Ivanitsky et al., 1981.) This wave pattern, as well as the target pattern, is not restricted to oscillatory systems but can also arise in a certain non-oscillatory version of the same reaction system. Unlike circular waves, no heterogeneities exist at the center of rotation. In order to cause rotation, a certain topological condition must be satisfied by the initial concentration distribution. The rotation may be clockwise or counterclockwise. It is common that the pattern arises as a clockwise-counterclockwise pair. A three-dimensional extension of rotating spiral waves is called scroll waves. If the chemical solution has a sufficient depth, the scroll axis is seen to close itself to form a scroll ring. In the argument below, we restrict ourselves to a two-dimensional pattern; scroll waves seem to be difficult to treat analytically except for the trivial case that the scroll axis forms a straight line.

From the success in obtaining target patterns analytically from a simple evolution equation for ϕ, one might be tempted to apply the same equation to spiral waves by seeking an angle-dependent solution of (6.3.4) without a U term, i.e.,

$$\frac{\partial Q}{\partial t} = \alpha(\beta\alpha^{-2} + \nabla^2)Q\,. \tag{6.5.1}$$

Unfortunately, however, such an attempt leads to a serious contradiction, the reason for which will be clarified below. The crucial point is that the spiral waves involve the notion of *phase singularity*. It is expected that the two-dimensional solution of (6.5.1) corresponding to a rotating spiral would be such that ϕ changes by an integer multiple of T each time the center of rotation (supposed to be situated at $r = 0$) was circled:

$$\phi(r, \theta + 2\pi, t) = \phi(r, \theta, t) + lT\,, \qquad l = \text{integer}\,. \tag{6.5.2}$$

Rotating waves with two and more arms have been observed by Agladze and Krinsky (1982) for the Belousov-Zhabotinsky reaction, and theoretically discussed by Koga (1982). We shall, however, restrict ourselves to the usual single-armed spiral waves for which $l = \pm 1$. As a further restriction, the pattern is assumed to rotate steadily (with frequency $\pm\Omega$, $\Omega > 0$), i.e.,

$$\phi(r, \theta, t) = \phi(r, \theta \pm \Omega t)\,. \tag{6.5.3}$$

Furthermore, ϕ is by definition an increasing function of t [remember that $d\phi/dt \simeq 1 + O(\varepsilon)$]. All the above requirements may be fulfilled by the form

$$\phi(r, \theta, t) = f(r) \pm (\theta \pm \Omega t)\,\omega_0^{-1}\,, \tag{6.5.4}$$

which we assume below, where $\omega_0 = 2\pi/T$ and $f(r)$ is some function of r. Then, Q must be of the form

$$Q(r, \theta, t) = \hat{Q}(r)\exp[\pm\beta\alpha^{-1}\omega_0^{-1}(\theta \pm \Omega t)]\,. \tag{6.5.5}$$

This is substituted into (6.5.1) to give the eigenvalue problem

$$\lambda\hat{Q}(r) = [\nabla_r^2 - U_0(r)]\,\hat{Q}(r)\,, \tag{6.5.6}$$

where $\nabla_r^2 \equiv d^2/dr^2 + r^{-1}d/dr$,

$$U_0(r) = -\left(\frac{p}{r}\right)^2, \quad p \equiv \frac{\beta}{\alpha\omega_0}\,, \tag{6.5.7}$$

and

$$\lambda = \beta\alpha^{-2}(\omega_0^{-1}\Omega - 1) . \tag{6.5.8}$$

Note that the attractive potential $U_0(r)$ arises entirely from the angular dependence of ϕ or Q, and not from heterogeneity. It may alternatively be said that the system is capable of producing effective pacemakers for itself. In contrast to the development of a target pattern, which was seen to be due to the entrainment by a real pacemaker, the entrainment by such a *virtual* pacemaker is considered to be the origin of the development of a spiral pattern.

In analogy to the case of circular waves, the contribution from the ground state is expected to be dominant. Therefore, λ will be understood hereafter to be the maximum eigenvalue. We now seek the asymptotic form of $\hat{Q}$ as $r \to \infty$. We have

$$\hat{Q}(r) \simeq e^{-\sqrt{\lambda}r} \tag{6.5.9}$$

in the lowest approximation. In general, one may assume the expansion

$$\hat{Q}(r) = e^{-\sqrt{\lambda}r} r^{-\nu}\left(1 + \frac{a}{r} + \frac{b}{r^2} + \dots\right) . \tag{6.5.10}$$

It is easy to confirm that the constants ν, a, b, ... can be determined iteratively. In particular, $\nu = 1/2$, so that (6.5.9) is improved to give

$$\hat{Q}(r) \simeq \exp(-\sqrt{\lambda}r - \tfrac{1}{2}\ln r) . \tag{6.5.11}$$

To this approximation,

$$\phi(r, \theta, t) = -\beta^{-1}\alpha\sqrt{\lambda}r - \frac{\alpha}{2\beta}\ln r \pm \omega_0^{-1}\theta + \omega_0^{-1}\Omega t . \tag{6.5.12}$$

Thus the contour of constant $\phi(r, \theta, t)$ is a spiral and coincides with the asymptotic form of an involute spiral as $r \to \infty$. Experimentally observed spiral waves also have this feature.

At this point a serious question arises as to whether the maximum eigenvalue λ is really finite. Actually, it cannot be finite. The reason is easily seen by rewriting (6.5.6) with the scaled coordinate $\tilde{r} \equiv \sqrt{\lambda}r$ as

$$\left(\nabla_{\tilde{r}}^2 + \frac{p^2}{\tilde{r}^2} - 1\right)\hat{Q}(\tilde{r}) = 0 . \tag{6.5.13}$$

Since λ does not appear explicitly in this equation, all the eigenfunctions should be represented by a universal function, and different eigenfunctions are simply interrelated through a scaling of the spatial coordinate. This clearly shows that λ (which must be positive) is unbounded. It should also be noted that as λ increases the corresponding eigenfunction becomes sharper and sharper in spatial variation, so that the assumption of a slow dependence on $\boldsymbol{r}$ on which our nonlinear phase diffusion equation is based becomes completely violated. In deriving this equation, the local composition vector X was assumed to deviate only slightly from the unperturbed limit cycle X_0. Under this restriction, it is absolutely impossible to imagine such a smooth continuous distribution of X near $r = 0$ as is compatible with the property (6.5.2).

6.6 Rotating Wave Solution of the Ginzburg-Landau Equation

The breakdown of our previous attempt to apply the phase description to spiral waves suggests that some simple model equations which allow for a strong orbital deformation would be valuable for the present purpose. In this connection, the Ginzburg-Landau equation appears to be most appropriate. As we see later, there is an interesting feature about this model, in that it can be transformed into an eigenvalue problem analogous to (6.5.6), except that the potential is strongly modified near the central core so that the ground state has a finite eigenvalue.

Before going into analytical theory, we show some results of the computer simulation carried out for the Ginzburg-Landau equation in the form of (2.4.18). From the nice symmetry of this model, we expect that the center of steady rotation is in the state of vanishing R, i.e., $(X, Y) = (0, 0)$, and hence the phase Θ of W cannot be defined there. Let this phase singularity be situated at $r = 0$ using polar coordinates (r, θ). Further, the rotation number l is assumed to be ± 1 as before:

$$l = \frac{1}{2\pi} \oint \nabla \Theta \cdot d\boldsymbol{r} = \pm 1 . \tag{6.6.1}$$

One may imagine a more general circumstance where a number of such phase singularities coexist in the system. Then the sum of the associated rotation numbers l_i must be conserved as long as none of them happen to be absorbed by the wall. A convenient initial distribution satisfying (6.6.1) is shown in Fig. 6.5 where X and Y have constant slopes in directions making 90° to each other. Consequently, the zero-level contours of X and Y intersect vertically at $r = 0$. It is clear that

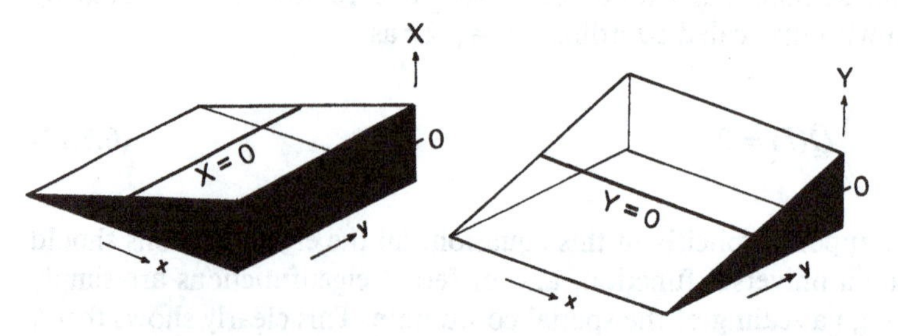

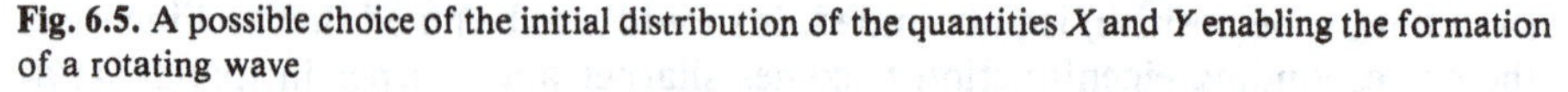
Fig. 6.5. A possible choice of the initial distribution of the quantities X and Y enabling the formation of a rotating wave

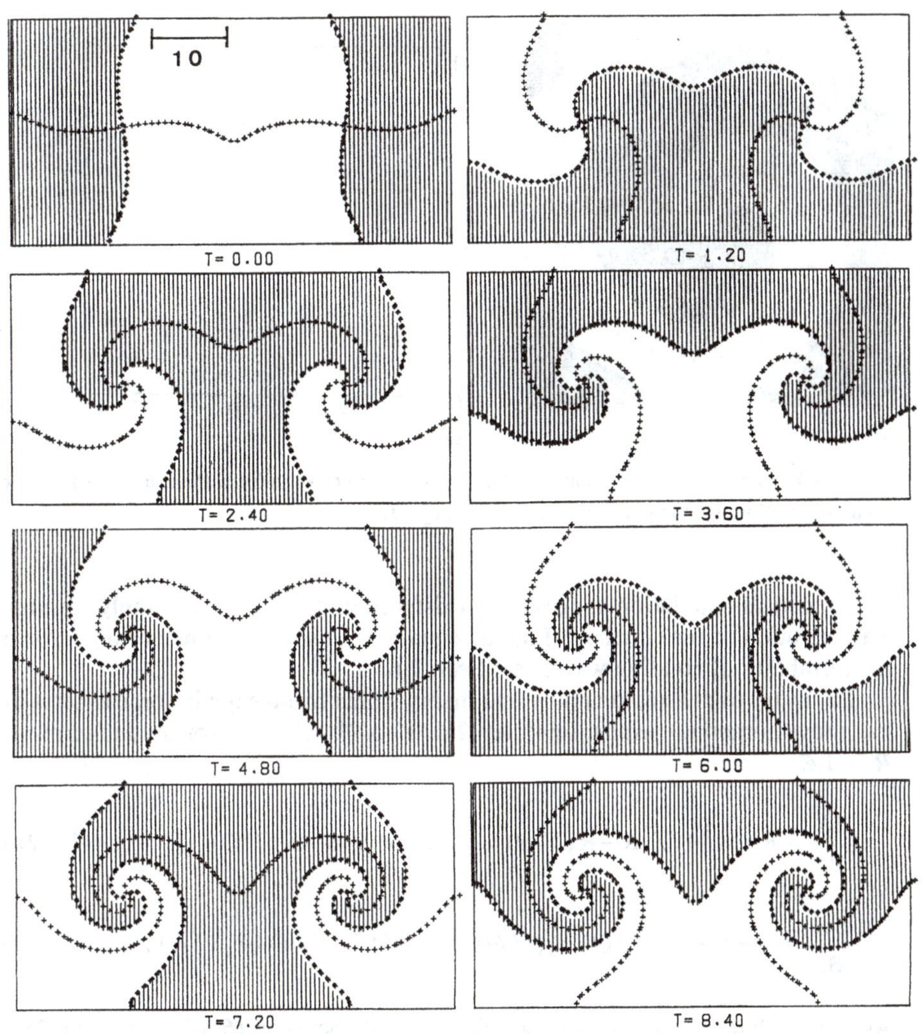

Fig. 6.6. Development of a pair of rotating waves. Shaded regions correspond to positive X. Contours of vanishing Y are also indicated for reference. Parameter values $c_1 = 0.0$, $c_2 = 1.8$

the rotation number about this phase singularity equals 1, as required. Since this number must be conserved, the system is unable to come back to the homogeneous oscillatory state even if the latter is a stable state. This kind of initial distribution for initiating rotation was employed by Winfree for a simple two-component excitable model (Winfree, 1974a). Assuming no-flux boundary conditions and with a pair of initial phase singularities, (2.4.18) was integrated numerically, and the evolution process of the pattern obtained is shown in Fig. 6.6. Clearly, the pattern is seen to approach a pair of steadily rotating spiral waves. For a single spiral wave in Fig. 6.7a, the amplitude R as a function of the distance from the center behaves like Fig. 6.7b. Although R was found to depend slightly on θ too, this comes only from the finiteness of the system size. In conclusion,

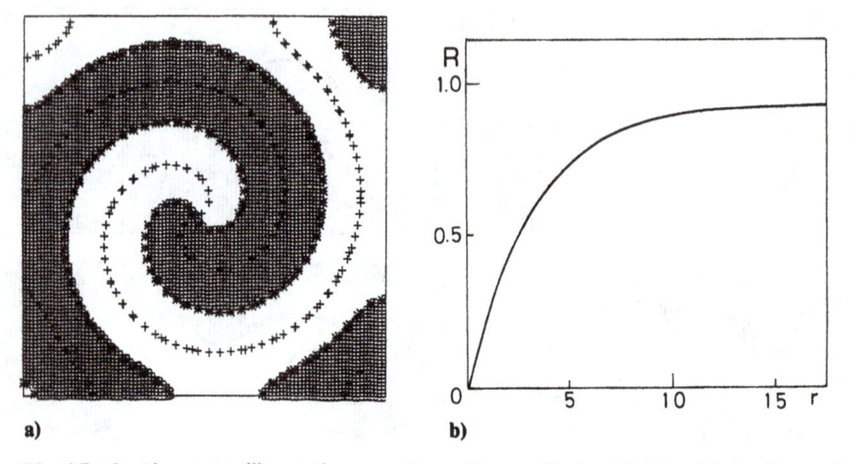

Fig. 6.7 a, b. Almost steadily rotating wave (**a**), and its amplitude variation with the distance from the center of rotation (**b**). Parameter values: $c_1 = 0.0$, $c_2 = 1.0$

the numerical simulation suggests the existence of a steadily and stably rotating wave solution of the Ginzburg-Landau equation for some parameter range. As to the instability of rotating waves, see Sect. 7.6.

Let us try to develop an analytical theory which could explain some aspects of the above numerical results. First, let (2.4.18) or (2.4.13) be expressed in terms of R and Θ:

$$\frac{\partial R}{\partial t} = R - R^3 + \nabla^2 R - R(\nabla \Theta)^2 - c_1(2 \nabla R \nabla \Theta + R \nabla^2 \Theta) , \tag{6.6.2a}$$

$$\frac{\partial \Theta}{\partial t} = c_0 - c_2 R^2 + \nabla^2 \Theta - c_1(\nabla \Theta)^2 + R^{-1}(2 \nabla R \nabla \Theta + c_1 \nabla^2 R) . \tag{6.6.2b}$$

Assuming that R is stationary, we eliminate R^2 between (6.6.2a) and (6.6.2b), and get

$$\frac{\partial \Theta}{\partial t} = \omega_0 + \alpha \nabla^2 \Theta + \omega_0^{-1} \beta (\nabla \Theta)^2 + 2 \alpha R^{-1} \nabla R \nabla \Theta - \omega_0^{-1} \beta R^{-1} \nabla^2 R , \tag{6.6.3}$$

where

$$\omega_0 = c_0 - c_2 ,$$

$$\alpha = 1 + c_1 c_2 ,$$

$$\beta = -(c_0 - c_2)(c_1 - c_2) .$$

Note that (6.6.3) reduces to the nonlinear phase diffusion equation if we neglect the space dependence of R. The previous numerical simulation suggests that R

has no angular dependence. Even with this simplification, solving the system of nonlinear equations (6.6.2a) (with $\partial R/\partial t = 0$) and (6.6.3) is a very formidable task. However, there is still hope of proceeding further if we notice the nonlinear transformation from Θ (or ϕ) to Q:

$$\begin{aligned}\Theta &= \omega_0\phi = \omega_0\beta^{-1}\alpha(\ln Q - \ln R)\\ &= p^{-1}(\ln Q - \ln R)\,,\end{aligned} \tag{6.6.4}$$

which is similar to (6.3.3). Substituting the above into (6.6.3) and noting that R is dependent only on r, we obtain

$$\frac{\partial Q}{\partial t} = \alpha[\beta\alpha^{-2} + \nabla^2 - (1+p^2)R^{-1}\nabla_r^2 R]Q\,. \tag{6.6.5}$$

This equation should be compared to (6.5.1). We are mainly interested in how the last term in (6.6.5) prevents the divergence of the maximum eigenvalue. The assumed form (6.5.5) for Q is now substituted into (6.6.5), which yields

$$\lambda\hat{Q}(r) = [\nabla_r^2 - U(r)]\hat{Q}(r)\,, \tag{6.6.6}$$

which should be compared to (6.5.6); here λ is given by (6.5.8) and

$$U(r) = U_0(r) + (1+p^2)R^{-1}\nabla_r^2 R\,. \tag{6.6.7}$$

We see that the space dependence of R results in an important modification in the potential. Equation (6.6.6) represents itself as a *nonlinear* eigenvalue problem, and it is generally difficult to solve it.

The only thing that should still be done analytically is to look into some specific properties of the modified potential. From the assumed form of Q in (6.5.5) together with (6.6.4), we see that the rotation $\theta \to \theta + \pi$ transforms Θ as $\Theta \to \Theta \pm \pi$, so that

$$W(r, \theta+\pi, t) = -W(r, \theta, t)\,. \tag{6.6.8}$$

In addition, W must be analytic at $r = 0$. From these facts, it follows that R and Θ are expanded near the point $r = 0$ as

$$R = a_1 r + a_3 r^3 + \dots\,, \tag{6.6.9a}$$

$$\Theta = \Omega t \pm \theta + d_2 r^2 + d_4 r^4 + \dots\,. \tag{6.6.9b}$$

On the other hand, the following expansions are expected far from the central core:

$$R = R_\infty + b_1 r^{-1} + b_2 r^{-2} + \dots\,, \tag{6.6.10a}$$

$$\Theta = \Omega t \pm \theta + f_{-1}r + f_0 \ln r + f_1 r^{-1} + f_2 r^{-2} + \dots\,. \tag{6.6.10b}$$

One may confirm that the expansion coefficients in (6.6.10a, b) can be determined from their substitution into (6.6.2a) (with $\partial R/\partial t = 0$) and (6.6.3). The asymptotic forms of U for small and large r can be known from the asymptotic forms of R in (6.6.9a) and (6.6.10a), respectively. We have

$$U(r) \to \frac{1}{r^2} + O\left(\frac{1}{r}\right) \quad \text{as} \quad r \to 0\,, \tag{6.6.11a}$$

$$\to U_0 + O\left(\frac{1}{r^3}\right) \quad \text{as} \quad r \to \infty\,. \tag{6.6.11b}$$

Thus the potential is made *repulsive* near the core in contrast to U_0, while it is attractive and approaches U_0 far from the core. Interestingly enough, $U(r)$ looks something like the interatomic potential. Since U remains essentially the same as U_0 for large r, the asymptotic wave pattern far from the core, where R is nearly constant, is again determined from (6.5.12), that is, an approximate involute spiral. The only new feature here is that the maximum eigenvalue λ is made finite by virtue of the strong repulsive part in the potential near the core.

7. Chemical Turbulence

Reaction-diffusion systems are expected to show spatio-temporal chaos in various circumstances. A few specific cases will be discussed. They include the turbulization of uniform oscillations, of propagating wave fronts and of rotating spiral waves.

7.1 Universal Diffusion-Induced Turbulence

People often speak of *chemical turbulence* whereby either of two distinct chaotic phenomena may be meant. One is the spatially uniform but temporally chaotic dynamics exhibited by the concentrations of chemical species, while the other involves spatial chaos too. For chemical turbulence in the latter sense, our attention is usually focused upon systems in which the local dynamics itself is non-chaotic, while such non-chaotic elements are coupled through diffussion to produce spatio-temporal chaos. In fact, if the local elements were already chaotic, the fields composed of them would trivially exhibit spatio-temporal chaos. Hence non-trivial chemical turbulence involving spatio-temporal chaos may be called *diffusion-induced chemical turbulence.*

In laboratory experiments, spatially uniform chemical turbulence (*chemical chaos* might be a better nomenclature) is generated in a well-stirred reactor. Usually, certain chemicals which are being consumed by reactions are fed into the reactor at a constant rate. Since diffusion plays no role in that case, the system may be mathematically modeled by a set of coupled ordinary differential equations whose dimension is the same as the number of the chemical species involved. There are good reasons why the rapidly growing interest in complicated behaviors of simple dynamical systems has been even more stimulated in recent years by chemical reactions, as a fascinating real example showing bifurcations and transitions to chaos. Most experimental work to date has been conducted on the Belousov-Zhabotinsky reaction. After some earlier experiments by Rössler and Wegman (1978), and by Schmitz et al. (1977), a fantastic bifurcation structure was discovered by Hudson, Hart and Marrinko (1979). For more recent works along the same line, we mention Vidal et al. (1980), Roux et al. (1981), Pomeau et al. (1981), Turner et al. (1981), and Simoyi et al. (1982). There is also an attempt (Nagashima, 1980) to find chaos in systems without flow. By virtue of these experimental studies together with some theoretical work (among others the elegant interpretation by Tomita and Tsuda, 1980), the subject of chemical chaos is becoming one of the most exciting topics of the chaotic dynamics of dissipative systems.

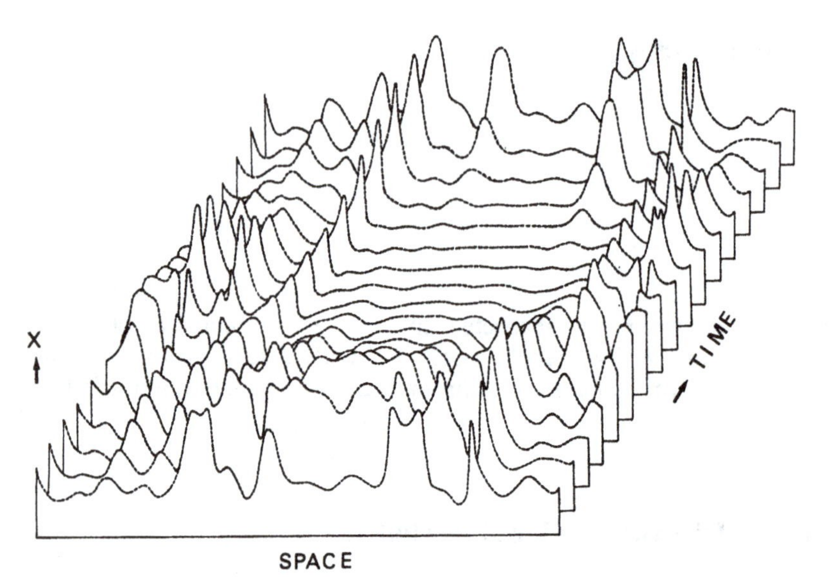

Fig. 7.1. Chaotic pattern of concentration X in the Brusselator in one space dimension, and its temporal development. Parameter values: $A = 2.0$, $B = 5.5$, $D_X = 1.0$, $D_Y = 0.0$ (for notation, see Appendix B)

Experiments and theories concerning the diffusion-induced type of chemical turbulence, to which this chapter is devoted, seem to have been only poorly developed so far, compared to the first type, although this is not surprising in view of the many difficulties involved therein. We do not even know at present of any definite real chemical examples of this type of turbulence, except for a suggestion of this possibility in the Belousov-Zhabotinsky system (Yamazaki et al., 1978, 1979). Still, the diffusion-induced turbulence could be ubiquitous in nature. Its possibilities in a variety of physical, chemical, and biological systems were vividly described by Rössler (1976, 1977). There is some mathematical work relating turbulence in reaction-diffusion systems to codimension two bifurcations (Kidachi, 1980; Guckenheimer, 1981). Theoretically, the condition for the occurrence of the diffusion-induced turbulence does not seem very restrictive. As a hypothetical chemical reaction model typically showing this behavior, one may mention the Brusselator (Kuramoto and Yamada, 1976a; Kuramoto, 1978). (For the Brusselator, see Appendix B.) Figure 7.1 shows a numerically simulated pattern of the concentration X of the Brusselator in one space dimension and its temporal evolution. Since the typical amplitude of the initial fluctuations about the uniform oscillations turned out to be less than 1/10 that of Fig. 7.1, the chaotic pattern in the figure may be viewed as genuine turbulence resulting from the instability of the uniform oscillations. We wish to know under what conditions such chaotic behavior is possible, not only for the Brusselator but also for more general reaction-diffusion systems, while we already know that perfectly coherent wave patterns can appear for the same class of systems under different conditions. It should also be clarified as to how the diffusion-induced chemical turbulence can be reduced, at least near its onset, to some known types of chaos such as those exhibited by systems of three coupled ordinary differential equations. Furthermore, there exist important statistical problems concerning the statistical nature of developed chemical turbulence, which will only be touched upon later.

The recent progress in the understanding of transitions to chaos in simple dynamical systems is remarkable. For continuous fields, i.e., systems with an infinite number of degrees of freedom, however, even the onset of chaos does not seem to have been understood to a satisfactory extent. As various experiments on fluid instabilities suggest (e.g., Fenstermacher et al., 1979; Gollub and Benson, 1980; Libchaber and Maurer, 1980, Giglio et al., 1981), there seem to exist no essential distinctions in the dynamical behavior between systems with an infinite number of degrees of freedom on the one hand, and systems with a very small number of degrees of freedom on the other hand, as far as the onset of chaos is concerned. This seems to be due to the feature common to many dissipative systems that most degrees of freedom except a few rapidly decay so that they are practically eliminated adiabatically (Haken, 1983a). Such a view should still be demonstrated explicitly, and this is a subject dealt with in the present chapter. For this purpose and for more general purposes of knowing the nature of chaos peculiar to continuous media, reaction-diffusion systems seem to have some advantages because of their mathematical simplicity compared, e.g., to the Navier-Stokes fluids. In particular, one space dimension is already sufficient to cause turbulence for reaction-diffusion systems as implied by the Brusselator turbulence shown above, and hence computer simulations of turbulence would be far easier to carry out.

The problems associated with chaos and turbulence in general are by no means restricted to their onset. How can latent degrees of freedom come up successively as some control parameter is varied, and how can we develop a satisfactory statistical dynamical description of such turbulence, involving a large number of *effective* degrees of freedom? The study of chemical turbulence might shed light on such statistical problems, although most of them will only find answers in the future.

We do not intend in this chapter to go beyond providing one of many possible approaches to diffusion-induced turbulence. The Ginzburg-Landau equation and the nonlinear phase diffusion equation (with an important modification) will play a principal role again. We will show later that the Ginzburg-Landau system can be turbulized (possibly after some bifurcations) if its uniform oscillation loses stability, although this does not represent the only possible origin of turbulence in this system. The instability of uniform oscillations in general reaction-diffusion systems occurs if α defined by (3.3.7) and (3.4.10b) becomes negative, which makes the nonlinear phase diffusion equation itself break down. As far as α remains only slightly negative, this equation can be modified by including the fourth derivative term $-\gamma\nabla^4\phi$ which causes damping, and hence counterbalances the instability term $\alpha\nabla^2\phi$. The equation may then be called the *phase turbulence equation*. We saw in Sect. 4.3 that the wavefront dynamics of some chemical waves is also described by nonlinear diffusion processes which are very similar to those for the phases of oscillating reaction-diffusion systems, see (4.3.28). Thus, the wavefronts should also be turbulized if α happens to be negative in the wavefront equation (4.3.28). Again, for small negative α, the phase turbulence equation is obtained. In this way, the phase turbulence equation appears in different physical contexts, so that the turbulence shown by it may have a universal nature. In the same sense, the turbulence shown by the Ginzburg-Landau equation may also represent a universal class of turbulence.

There may be an additional value in studying spatio-temporal chemical turbulence, in connection with its possible relevance to some biological problems. This is expected from the fact that the fields of coupled limit cycle oscillators (or non-oscillating elements with latent oscillatory nature) are often met in living systems. In some cases, such systems show orderly wavelike activities much the same as those observed in the Belousov-Zhabotinsky reaction. There seems to be no reason why we should not expect such organized motion to become unstable and hence show turbulent behavior. The recent work by Ermentrout (1982) who derived a Ginzburg-Landau type equation for neural field seems to be of particular interest in this connection.

7.2 Phase Turbulence Equation

The stability problem of the uniform time-periodic solution $X_0(t)$ to the reaction-diffusion equations is formally developed as follows. Here the system size is assumed to be infinitely large. Let the deviation $\boldsymbol{u}(t)$ about $X_0(t)$ be expressed as a Fourier series:

$$\boldsymbol{u}(\boldsymbol{r},t) = \sum_{q} \boldsymbol{u}_q(t)\mathrm{e}^{\mathrm{i}qr}\,.$$

Since $\boldsymbol{u}$ is real, $\boldsymbol{u}_q = \bar{\boldsymbol{u}}_{-q}$. The reaction-diffusion equations are then linearized in $\boldsymbol{u}_q$:

$$\frac{\partial \boldsymbol{u}_q}{\partial t} = L_q(t)\boldsymbol{u}_q(t)\,, \quad \text{where} \tag{7.2.1}$$

$$L_q(t) = L(t) - Dq^2\,. \tag{7.2.2}$$

This equation is a simple generalization of (3.4.1). Analogously to (3.4.2), the general solution of (7.2.1) becomes

$$\boldsymbol{u}_q(t) = S_q(t)\mathrm{e}^{\Lambda_q t}\boldsymbol{u}_q(0)\,, \tag{7.2.3}$$

where $S_q(t+T) = S_q(t)$, $S_q(0) = 1$ and

$$\frac{dS_q(t)}{dt} + S_q(t)\Lambda_q - L_q(t)S_q(t) = 0\,. \tag{7.2.4}$$

The stability of the uniform oscillations to small-amplitude non-uniform fluctuations is determined from the eigenvalues of Λ_q. The eigenvectors $\boldsymbol{u}_l$ and $\boldsymbol{u}_l^*$ in Sect. 3.4 are now generalized to include dependence on the wavevector q. Thus,

$$\Lambda_q \boldsymbol{u}_{lq} = \lambda_{lq}\boldsymbol{u}_{lq}\,, \quad \boldsymbol{u}_{lq}^*\Lambda_q = \lambda_{lq}\boldsymbol{u}_{lq}^*\,, \quad l = 0,1,\ldots,n-1\,, \tag{7.2.5}$$

where $u_{lq}^* u_{mq'} = \delta_{lm}\delta_{qq'}$ are assumed. The n eigenvalues λ_l for the uniform system are then extended to form n branches. The lowest branch $l = 0$ may be called the phaselike branch and the others the amplitudelike branches. Let us assume the asymptotic orbital stability of $X_0(t)$ to *uniform* fluctuations, which means that $\mathrm{Re}\{\lambda_l\} < 0$ for $l \neq 0$. For sufficiently small q, the eigenvalues λ_{lq} may be expanded as

$$\lambda_{lq} = \lambda_l + \lambda_l^{(1)} q^2 + \lambda_l^{(2)} q^4 + \dots \, . \tag{7.2.6}$$

It is thus obvious that the stability of the uniform oscillations to long-wavelength fluctuations depends crucially on the sign of $\mathrm{Re}\{\lambda_0^{(1)}\}$; a negative sign implies stability, and a positive sign instability. We wish to obtain the expansion coefficients in (7.2.6), in particular $\lambda_l^{(1)}$, in terms of Λ, S, and D. To do this, we develop Λ_q and u_{lq} as

$$\Lambda_q = \Lambda + q^2 \Lambda^{(1)} + q^4 \Lambda^{(2)} + \dots \, , \tag{7.2.7}$$

$$u_{lq} = u_l + q^2 u_l^{(1)} + q^4 u_l^{(2)} + \dots \, . \tag{7.2.8}$$

Substituting (7.2.6 – 8) into (7.2.5), and equating coefficients of different powers of q^2, we obtain

$$\lambda_l^{(1)} = (l|\Lambda^{(1)}|l) \, , \tag{7.2.9}$$

$$\lambda_l^{(2)} = \sum_{m \neq l} (l|\Lambda^{(1)}|m)(m|\Lambda^{(1)}|l)/(\lambda_l - \lambda_m) + (l|\Lambda^{(2)}|l) \, , \tag{7.2.10}$$

etc. Now the problem reduces to finding $\Lambda^{(1)}$, $\Lambda^{(2)}$, etc., in terms of Λ, S, and D. Substituting (7.2.2, 7) and also the expansion

$$S_q = S(1 + q^2 s^{(1)} + q^4 s^{(2)} + \dots)$$

into (7.2.4), we have

$$\frac{dS(t)}{dt} + S(t)\Lambda - L(t)S(t) = 0 \, , \tag{7.2.11 a}$$

$$\frac{ds^{(1)}(t)}{dt} + [s^{(1)}(t), \Lambda] + \Lambda^{(1)} + S^{-1}(t)DS(t) = 0 \, , \tag{7.2.11 b}$$

$$\frac{ds^{(2)}(t)}{dt} + [s^{(2)}(t), \Lambda] + \Lambda^{(2)} + s^{(1)}(t)\Lambda^{(1)} + S^{-1}(t)DS(t) \cdot s^{(1)}(t) = 0 \, , \tag{7.2.11 c}$$

....

where $[A, B] \equiv AB - BA$. Since the $s^{(i)}(t)$ are T-periodic in t, then (7.2.11 b, c), etc., are time averaged to give

$$\Lambda^{(1)} = -\frac{1}{T}\int_0^T dt\{[s^{(1)}(t), \Lambda] + S^{-1}(t)DS(t)\}, \tag{7.2.12a}$$

$$\Lambda^{(2)} = -\frac{1}{T}\int_0^T dt\{[s^{(2)}(t), \Lambda] + s^{(1)}(t)\Lambda^{(1)} + S^{-1}(t)DS(t)\cdot s^{(1)}(t)\}, \tag{7.2.12b}$$

....

Thus, it immediately follows that

$$\lambda_l^{(1)} = -\frac{1}{T}\int_0^T (l|S^{-1}(t)DS(t)|l)\,dt. \tag{7.2.13}$$

The higher-order coefficients $\lambda_l^{(2)}$, etc., may also be calculated iteratively, but the calculation soon becomes very cumbersome. By comparison of (7.2.13) for $l = 0$ with (3.3.7) combined with (3.4.10b), we find

$$\lambda_0^{(1)} = -\alpha. \tag{7.2.14}$$

This shows the anticipated fact that the instability of the uniform oscillation to long wavelength fluctuations corresponds precisely to the negative sign of the phase diffusion constant. The equality $\lambda_0^{(2)} = -\gamma$ may also be confirmed, where γ is the quantity which appeared in (4.2.36) and is the abbreviation of $-\omega_2^{(1)}$ defined in (4.2.35). More generally, it is possible to prove that the dispersion curve of the phaselike branch has an exact correspondence to the linearized form of the phase diffusion equation (4.2.36), or one may possibly have

$$\frac{\partial\psi}{\partial t} = -\lambda_0^{(1)}\nabla^2\psi + \lambda_0^{(2)}\nabla^4\psi - \lambda_0^{(3)}\nabla^6\psi + \dots .$$

Suppose α turned out to be negative. Then the dispersion curves would look like Fig. 7.2, where $\gamma > 0$ is assumed. If α is only slightly negative as in Fig. 7.2a, the unstable phaselike fluctuations are expected to be characterized by a very long time scale t_c and a very long space scale r_c. These characteristic scales may be estimated on the assumption that the effect of the instability, represented by the $\alpha\nabla^2\psi$ term, and that of the damping, $\gamma\nabla^4\psi$, and also the term $\partial\psi/\partial t$ are comparable in magnitude, or

$$\alpha r_c^{-2} \sim \gamma r_c^{-4} \sim t_c^{-1}, \quad \text{or}$$

$$t_c \sim |\alpha|^{-2}, \quad r_c \sim |\alpha|^{-1/2}, \tag{7.2.15}$$

where γ has been assumed to be $O(1)$ and not included in the last expression. For such weak phase instability, the amplitudelike fluctuations have far shorter time scales, so that they are expected to follow adiabatically the slow motion of the unstable phaselike fluctuations. This suggests that we are allowed to employ the phase description of Chap. 4. In contrast, when the phase instability is relatively

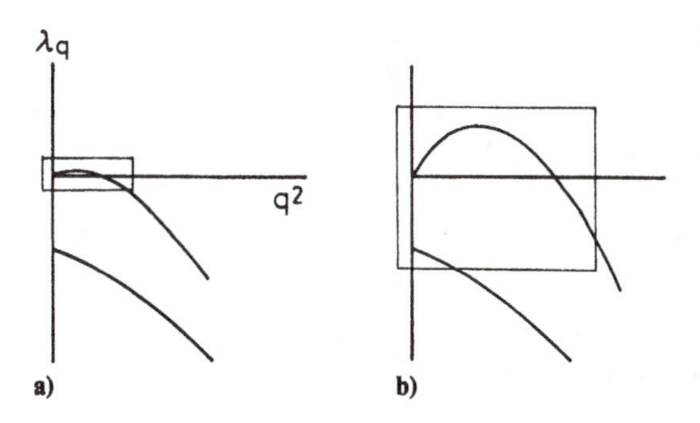

Fig. 7.2. (a) Weak phase instability and (b) strong phase instability. The characteristic time scales of unstable phase fluctuations and of stable amplitude fluctuations are clearly separated for (a) but not for (b)

strong (Fig. 7.2b), the unstable phaselike fluctuations may have time scales comparable to those of some amplitude fluctuations, so that the phase description is no longer valid. This latter case will be taken up in Sect. 7.5; our concern below in this section is restricted to small $|\alpha|$.

We obtained in Chap. 4 a general expansion for $d\phi/dt$ (or $d\psi/dt$) in the form of (4.2.36), but the question is which terms in the expansion should be retained. The point here is that ∇ no longer represents a unique small parameter because of the presence of an additional small parameter $|\alpha|$. However, the smallness associated with ∇ cannot not be independent of the smallness of $|\alpha|$; in fact, the second relation in (7.2.15) implies (symbolically)

$$\nabla \sim |\alpha|^{1/2}. \tag{7.2.16}$$

Furthermore, ψ itself is considered to have a certain characteristic amplitude of variation depending on $|\alpha|$. Taking all these facts into consideration, we see that a convenient way to single out leading terms from (4.2.36) is to postulate a scaling form for ψ as

$$\psi(r,t) = |\alpha|^{\lambda}\tilde{\psi}(|\alpha|^{\mu}t, |\alpha|^{\nu}r), \tag{7.2.17}$$

where λ, μ, and ν are supposed to be positive constants. The relations in (7.2.15) imply that $\mu = 2$ and $\nu = 1/2$, but the same results and also the value of λ are obtained from a more natural argument as follows. Let (7.2.17) be substituted into (4.2.36), which leads to

$$\begin{aligned}\frac{\partial\tilde{\psi}}{\partial\tau} &= \alpha|\alpha|^{2\nu-\mu}\nabla_{\tilde{r}}^{2}\tilde{\psi} + \beta|\alpha|^{\lambda+2\nu-\mu}(\nabla_{\tilde{r}}\tilde{\psi})^{2}\\ &\quad - \gamma|\alpha|^{4\nu-\mu}\nabla_{\tilde{r}}^{4}\tilde{\psi} + \dots,\end{aligned} \tag{7.2.18}$$

where τ and $\tilde{r}$ are the scaled time and space coordinates,

$$\tau = |\alpha|^{\mu}t, \qquad \tilde{r} = |\alpha|^{\nu}r.$$

It is seen that this choice of the μ and ν values, together with the additional assumption $\lambda = 1$, makes the term $\partial\tilde{\psi}/\partial\tau$ and the first three terms on the right-hand side of (7.2.18) balance each other in magnitude, and makes all the other terms, i.e., terms not explicitly written in (7.2.18), negligible as $|\alpha| \to 0$. Furthermore, there does not seem to exist any other choice of indices which is physically meaningful. Thus, to the lowest order in $|\alpha|$, we have

$$\frac{\partial \psi}{\partial t} = \alpha \nabla^2 \psi + \beta (\nabla \psi)^2 - \gamma \nabla^4 \psi . \tag{7.2.19}$$

In what follows, we always assume $\gamma > 0$. The solution of (7.2.19) turns out chaotic for sufficiently large system size, and will be analyzed in Sect. 7.4. This equation may be called the phase turbulence equation. Recently, the same partial differential equation was derived by Sivashinsky in connection with the dynamics of combustion, and was used in discussing the turbulization of flame fronts (Sivashinsky, 1977, 1979; Michelson and Sivashinsky, 1977).

As shown in Appendix B, α can happen to be negative for the Brusselator, at least in the vicinity of the Hopf bifurcation point. Moreover, it is seen from (B.19) that $|\alpha|$ can be made arbitrarily small by suitably choosing A, D_X, and D_Y, while β and γ can remain of ordinary magnitude. Thus, we have at least one chemical reaction model showing phase turbulence.

It would be instructive here to make an intuitive argument to get some insight into the mechanism behind the instability of the uniform oscillations. In this connection, the instability condition

$$1 + c_1 c_2 < 0 \tag{7.2.20}$$

(Appendix A) for the Ginzburg-Landau system is very suggestive; this inequality may be viewed as a condensed expression of the universal mechanism of the phase instability in oscillatory reaction-diffusion systems not restricted to the vicinity of the Hopf bifurcation point. Remember that c_1 measures the degree to which the diffusion matrix of the starting reaction-diffusion equations deviates from a scalar. On the other hand, c_2 represents how strongly the frequencies of the individual local oscillators depend on their amplitudes; this is obvious if one expresses (2.4.15) as

$$\frac{\partial W}{\partial t} = (1 - |W|^2 + \mathrm{i}\tilde{\omega}) W + (1 + \mathrm{i}c_1) \nabla^2 W , \tag{7.2.21}$$

where the *effective* frequency $\tilde{\omega}$ is given by

$$\tilde{\omega} = -c_2 |W|^2 . \tag{7.2.22}$$

The instability seems to result from the cooperation of the two effects each represented by c_1 and c_2. In Fig. 7.3 (see also Fig. 4.3c) we have shown schematically a one-dimensional array of limit cycle oscillators, each in a two-dimensional state space (i.e., complex W space). The line formed by joining the local oscillator states may be imagined as an elastic string circulating round a cylinder surface. Suppose that initially the string was perfectly uniform

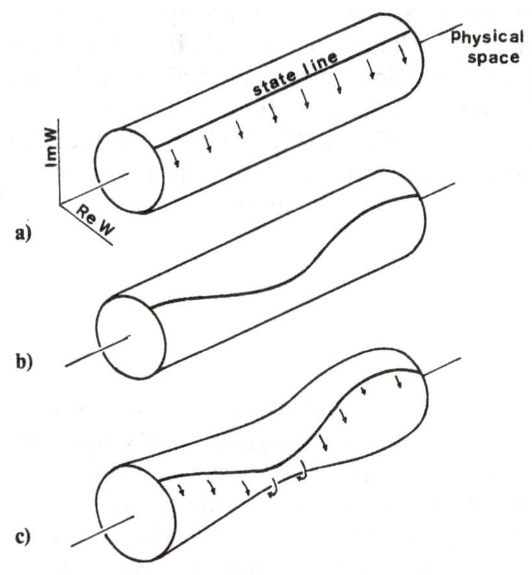

Fig. 7.3a–c. Geometrical interpretation of limit cycle oscillations with uniform phase (**a**) and non-uniform phase (**b**). The phase non-uniformity generally induces some non-uniformity in local frequencies, which may cause instability (**c**)

(Fig. 7.3a). We now disturb the state line slightly by giving a wavy *phase* modulation like Fig. 7.3b. Note that the string still lies on the cyclinder surface. How does this phase disturbance develop thereafter? The diffusion may have a tendency to even out the wavy modulation. This obvious effect is reflected in the inequality (7.2.20) by the term 1, and favors the stability of the uniform oscillation. The second effect represented by the term $c_1 c_2$ is a little puzzling. The imaginary part c_1 of the diffusion constant has the effect of rotating the state line round itself. As a result the state line must necessarily deviate from the original cylinder surface; in some parts it may go inward, and in other parts outward. In this way, the c_1 term has the effect of transforming the phase fluctuations into amplitude fluctuations. Since the frequency of the local oscillators was seen to depend on their amplitudes (due to the presence of the c_2 term), such a wavy amplitude profile implies a similar profile of the local frequencies. For suitable signs of c_1 and c_2, it may happen that the phase-advanced parts of the state line find increases in their local frequencies, while for the phase-retarded parts, the frequencies will be lowered. If such an effect is strong enough to cancel the stabilizing effect mentioned above, then the wavy phase modulation will become even stronger (Fig. 7.3c), which leads to instability.

The possibility of "diffusion instability" in reaction-diffusion systems has long been known since the work by Turing (1952) and even traces back to Rashevsky (1940). Some people argue that it is a key mechanism in the formation of some ecological patterns (Segel and Jackson, 1972) or in morphogenesis (Gierer and Meinhardt, 1972). The diffusion instability of this kind and our phase instability are different things in that the system state which is being destabilized is an equilibrium state in one case and an oscillating state in the other case. It may still be suspected that the phase instability might be a version of this traditional diffusion instability, or a disguised form, simply due to the presence of oscillations. This view turns out incorrect, however, by taking the Brusselator, for example. This model can exhibit the usual diffusion instability as well as the present type of phase instability, which is explained in Appendix B. Not far from the Hopf bifurcation point, a necessary condition for the conventional diffusion instability is found to be $D_X < D_Y$ [compare (B.7, 8)], while for the occurrence of phase instability, we must have $D_X > D_Y$, see (B.18, 19). However, the fact that these two types of diffusion instabilities should be mutually exclusive has never been proved in the general context of reaction and diffusion.

7.3 Wavefront Instability

We derived in Chap. 4 an evolution equation for slowly varying wavefronts in two-dimensional reaction-diffusion systems. Quite analogously to the dynamics of oscillatory systems with a slowly varying phase pattern, we obtained an asymptotic expansion (4.3.28). If α happened to be small and negative, while the other parameters were of ordinary magnitude and γ positive, then the same reasoning advanced in Sect. 7.2 applies, and we get the one-dimensional phase turbulence equation

$$\frac{\partial \psi}{\partial t} = \alpha \frac{\partial^2 \psi}{\partial y^2} + \beta \left(\frac{\partial \psi}{\partial y} \right)^2 - \gamma \frac{\partial^4 \psi}{\partial y^4} . \tag{7.3.1}$$

One may wonder if there exist any specific reaction-diffusion models giving rise to negative α. Such a model in fact exists, for which one may even prove analytically the possibility of arbitrarily small $|\alpha|$. This is a piecewise linear version of the Bonhoeffer – van der Pol model including diffusion, and is given by

$$\begin{aligned} \frac{\partial X}{\partial t} &= -X + H(X-a) - Y + D_X \nabla^2 X , \\ \frac{\partial Y}{\partial t} &= bX - cY + D_Y \nabla^2 Y . \end{aligned} \tag{7.3.2}$$

Here H is the step function, i.e., $H(\alpha) = 1$ or 0 according to $\alpha > 0$ or $\alpha < 0$, and a, b, c, D_X, and D_Y are non-negative constants. For some special parameter values, the above model is sometimes employed for studying pulse propagation in nervelike excitable systems or the dynamics of rotating waves. The work by McKean (1970), and by Rinzel and Keller (1973) concerns the case of vanishing c and D_Y. The model may then be viewed as an idealization of the FitzHugh-Nagumo nerve conduction equation (FitzHugh, 1961; Nagumo et al., 1962). Rinzel and Keller obtained analytic solutions for pulse propagation and analyzed their stability. Winfree (1978) developed an interesting intuitive argument about rotating waves putting emphasis on the curious nature of the phase singularity involved, and demonstrated his idea by making a numerical simulation for the case with equal diffusion constants and vanishing c. The potential richness

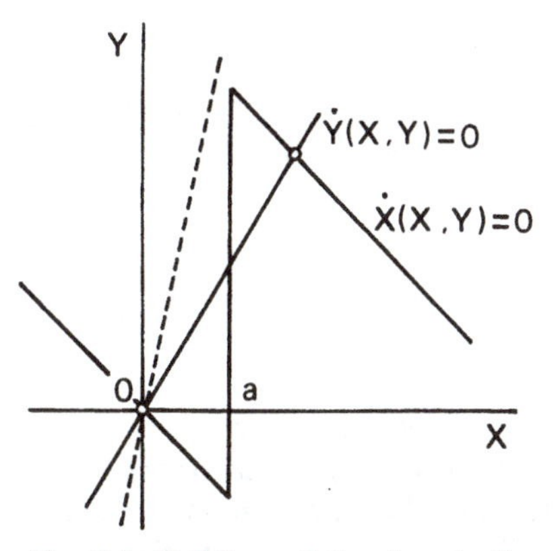

Fig. 7.4. Nullclines of the piecewise-linear Bonhoeffer – van der Pol model. For larger b/c, the system becomes monostable as indicated by the broken line

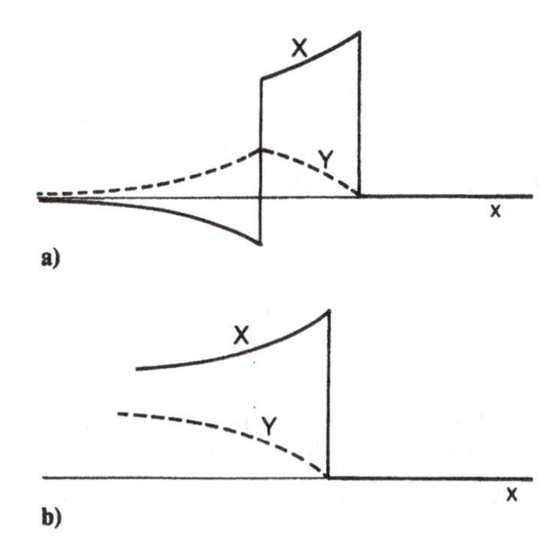

Fig. 7.5 a, b. A pulse (**a**) and kink (**b**) with sharp wavefronts

of the model (7.3.2) seems to be unexpectedly great in contrast to its simple appearance.

Some qualitative features of the diffusionless part of the model (7.3.2) are now pointed out. The nullclines [i.e., the curves in the XY plane obtained by equating the respective right-hand sides of (7.3.2) to zero] are shown in Fig. 7.4. The first nullcline is a caricature of sigmoidal manifolds such as appear in many differential-equation models for the kinetics of excitable membranes, enzyme reactions, etc. (see, e.g., Murray, 1977). The second nullcline is simply a straight line. The intersection points of these lines give steady states, so that one may have one or two stable steady states depending on the parameter values. In the case of one steady state, the system possesses an excitable nature. Then, by including diffusion, one may expect the appearance of pulses or trigger waves as schematically indicated in Fig. 4.2a under suitable local initial stimuli. In the case of bistability, the system including diffusion can produce interfaces which connect smoothly the different steady states (Fig. 4.2b). Below we will give some analytical results concerning the front diffusion constant α (for more details, see Kuramoto, 1980a, b). It can be shown that the problem of calculating α is greatly simplified if b and c are very small [they are supposed to be $O(\varepsilon)$]. Then the pulse and kink obtained look like Figs. 7.5a and b, respectively, in a suitably stretched space scale. One obvious feature here is the existence of two distinct characteristic length scales, one corresponding to the very sharp transition region, and the other to the remaining much smoother part. It is known that a singular perturbation method is available for such waves (Ortoleva and Ross, 1975; Fife, 1976a, b). Further analysis shows that sharp wavefronts such as those in Fig. 7.5 are very stable to *lateral* distortions as long as the ratio D_Y/D_X is not too large. In the usual nerve conduction problems, D_Y is taken to be zero, which implies perfect stability. The other extreme, i.e., $D_Y/D_X \to \infty$ as $\varepsilon \to 0$, seems to be of interest in relation to the front instability. When $D_X/D_Y = O(\sqrt{\varepsilon})$, in particular, α can show transitions between negative and positive values. To be specific, let us put

$$b = \varepsilon\tilde{b}\,, \quad c = \varepsilon\tilde{c}\,, \quad D_X = \varepsilon^{1/2}\tilde{D}_X\,,$$

and suppose the quantities with a tilde as well as a and D_Y to be of $O(1)$. Then the front diffusion constant α can be calculated to the lowest order in ε to give (Kuramoto, 1980a)

$$\alpha = \varepsilon^{1/2}\tilde{D}_X\left[1 - \frac{\tilde{b}}{(1-2a)^2}\left(\frac{D_Y}{\tilde{D}_X}\right)^2\right]. \tag{7.3.3}$$

Clearly, α can be positive or negative depending on the parameter values.

In contrast to the instability of uniform oscillations, the physical origin of the wavefront instability seems to have some relation to the conventional diffusion instability. To see this, we first give a brief qualitative interpretation of the conventional diffusion instability. Here the notions *activator* and *inhibitor* seem to be helpful, and for simplicity we imagine a two-component activator-inhibitor system. The instability then turns out to be due to relatively rapid diffusion of the inhibiting substance. Consider the activator-inhibitor kinetics (first, without diffusion):

$$\frac{dX}{dt} = f(X, Y)\,, \quad \frac{dY}{dt} = g(X, Y)\,. \tag{7.3.4}$$

The shapes of the nullclines $f = 0$ and $g = 0$ are supposed to be such as shown in Fig. 7.6a. Let ξ and η denote the deviations from the intersection point (X_0, Y_0). Linearization of (7.3.4) about this point leads to

$$\frac{d\xi}{dt} = \alpha\xi - \beta\eta\,, \quad \frac{d\eta}{dt} = \gamma\xi - \delta\eta\,. \tag{7.3.5}$$

Since the slopes of the nullclines at (X_0, Y_0) are both positive, we have

$$\alpha, \beta, \gamma, \delta > 0\,, \tag{7.3.6a}$$

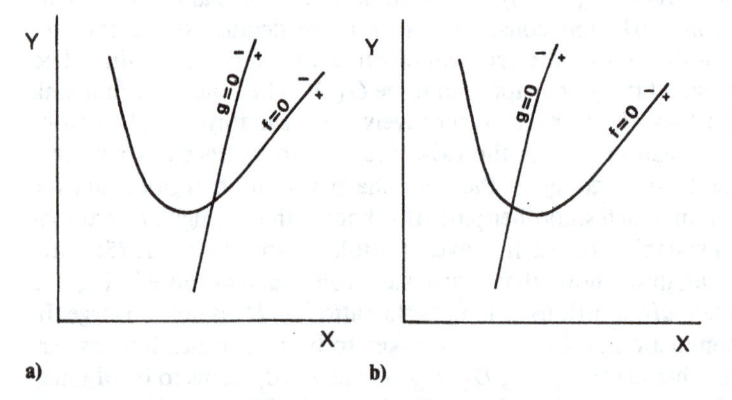

Fig. 7.6a, b. Two typical intersections between the nullclines in a two-component system represented by (7.3.4)

and the figure clearly shows

$$\alpha/\beta < \gamma/\delta . \tag{7.3.6b}$$

Under these conditions, the stability of the steady state is guaranteed if

$$\delta > \alpha , \tag{7.3.7}$$

the violation of which leads to oscillatory instability. Let the steady point lie close enough to the minimum of the nullcline $f = 0$ so that the condition (7.3.7) may be well satisfied. Then the way ξ and η behave is such that these species may be suitably called the activator and inhibitor, respectively. In fact ξ has its own tendency to blow up autocatalytically due to the term $\alpha\xi$, whereas η has the nature of suppressing the growth of ξ through the term $-\beta\eta$.

With this picture in mind, the stability of the steady state may be qualitatively accounted for as follows. Let the activator concentration deviate slightly from its stationary value X_0. This deviation tends to grow exponentially as noted, but its occurrence immediately causes the production of the inhibitor through the term $\gamma\xi$. The resulting excess inhibitor concentration is then used to suppress the increase of the activator concentration through the term $-\beta\eta$, thus forcing X back to its original stationary value. According to the intrinsic stable nature of the inhibitor as represented by the term $-\delta\eta$, Y also comes back to its stationary value. The stabilization mechanism like this can operate for arbitrarily small deviations, which means that no spontaneous departure from the steady state can actually occur. It would be interesting to examine how such a picture of stability has to be modified when the diffusion terms $D_X\partial^2\xi/\partial x^2$ and $D_Y\partial^2\eta/\partial x^2$ are included in the respective equations in (7.3.5). Formal stability analysis is almost trivial. By assuming the space-time dependence for ξ and η to be $\exp(\lambda t + \mathrm{i}kx)$, we get the condition

$$\frac{\alpha - D_X k^2}{\beta} < \frac{\gamma}{\delta + D_Y k^2} \tag{7.3.8}$$

for the stability of the uniform steady state to non-uniform fluctuations with wavenumber k. Note that the above condition is a generalization of (7.3.6b) and the former can be violated for non-vanishing k even if it is satisfied for vanishing k. The instability to non-vanishing k is nothing but the Rashevsky-Turing or the conventional diffusion instability. This occurs in particular for D_X sufficiently small and D_Y sufficiently large. Why the fast inhibitor diffusion (or slow activator diffusion) causes instability may be interpreted as follows. Let X be locally perturbed from its equilibrium value. As before, the fluctuation thus produced has its own tendency to be amplified exponentially as far as it does not diffuse out too rapidly. Now the inhibitor experiences this local fluctuation of the activator, so that its concentration there is made somewhat higher than its equilibrium value. Since the diffusion of the inhibitor is assumed to be fast, this excess amount of Y soon diffuses out. Consequently, the local inhibitor concentration can no longer be high enough to suppress the autocatalytic growth of the activator concentration, which implies instability.

So much for the interpretation of the usual diffusion instability. We now let the nullclines $f=0$ and $g=0$ intersect in a slightly different way as indicated in Fig. 7.6b, where the line $f=0$ has a negative slope at the intersection point in contrast to the previous situation. If we write the linearized equation in the form of (7.3.5), we have

$$\alpha<0, \quad \beta, \gamma, \delta>0.$$

Note that our piecewise linear model corresponds qualitatively to this situation. The condition (7.3.8) is automatically satisfied for all k, and the steady state is stable irrespective of the presence or absence of diffusion. Still, the system may exhibit pulses or kinks under suitable initial conditions. Although such systems are not usually called activator-inhibitor systems, they still retain some similarity to activator-inhibitor systems if the flow in the XY phase space is seen globally beyond the linear regime about the steady state. In fact this similarity to activator-inhibitor systems has some connection with the similarity of the front instability to the conventional diffusion instability. Suppose that $a \ll 1$. If the equilibrium value of X (i.e., the zero value) is perturbed slightly but beyond the small threshold value a, then we have $\partial X/\partial t \simeq 1$, so that X starts to grow (though not exponentially). The increase of X causes the "inhibitor" Y to be produced via the term bX, which in turn suppresses the growth of X via the term $-Y$. If this suppression by Y is not strong enough (i.e., if c is relatively small), the system will reach a different steady state where both the "activator" and "inhibitor" are richer than before. This occurs under bistability condition. For monostable cases, X and Y will ultimately come back to the original steady values, but this is only possible after a rather long excursion in XY phase space.

Now the origin of the wavefront instability (at least for our piecewise linear system, and possibly for wider classes of systems) may be interpreted as follows. For definiteness, we restrict consideration to bistable systems, and the same reasoning may be carried over to monostable excitable systems. Initially, the medium is supposed to be partitioned into two distinct regions corresponding to different steady states, the interface or wavefront being uniform and moving to the right (Fig. 7.7a). Assume the steady state in the right domain has lower "activator" and "inhibitor" concentrations, while they are higher in the left domain. (Interchange of "right" and "left" in the last statement does not alter the conclusion.) Let the front be slightly distorted as in Fig. 7.7b. Since Y diffuses rapidly, whereas X diffuses slowly, closer observation would reveal that the wavefront defined by some isoconcentration contour of Y is somewhat smoother than that defined through X. As a consequence, the most advanced part of the front (indicated as A in the figure) finds a poorer "inhibitor" compared to the case with no front distortion; conversely, the neighboring parts B and C find a richer "inhibitor". The deficiency in Y accelerates the production of X while excessive Y decelerates it, which means that the local propagation speed is increased near A and decreased near B and C. If such an effect is strong enough to exceed the ordinary smoothening effect which always exists, then the front becomes

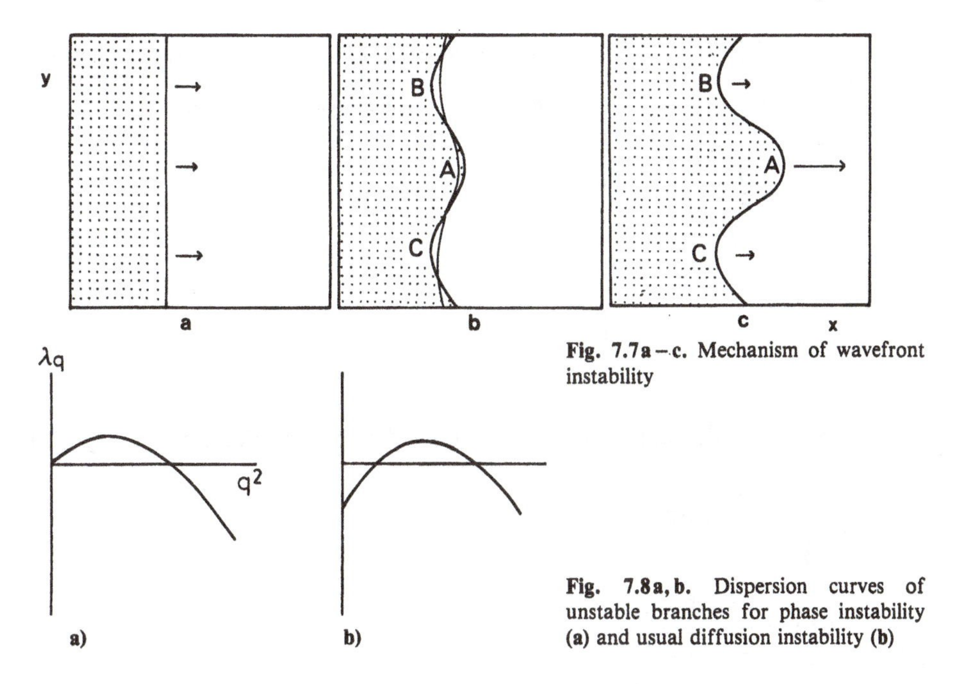

Fig. 7.7a–c. Mechanism of wavefront instability

Fig. 7.8a, b. Dispersion curves of unstable branches for phase instability (**a**) and usual diffusion instability (**b**)

even more distorted, and this leads to instability. The usual diffusion instability gives rise to ordered spatial structures, while the wavefront instability easily leads to chaotic patterns. This distinction is possibly related to the difference in their linear dispersion characteristics as contrasted in Fig. 7.8.

Although some physical implications of the nonlinear phase diffusion equation with positive α have been discussed in Sect. 6.2, we have not yet discussed the same equation in relation to the wavefront dynamics; this should be done before going into the phase turbulence equation. Let the wavefront form a straight line which is slightly non-parallel to the y direction (Fig. 7.9). Then the nonlinear phase diffusion equation becomes

$$\frac{\partial \psi}{\partial t} = \beta \left(\frac{\partial \psi}{\partial y} \right)^2 . \tag{7.3.9}$$

Remember that the x coordinate of the front which we denote by $x_f(y, t)$ is related to ψ by

$$x_f(y, t) = c[t + \psi(y, t)] , \tag{7.3.10}$$

see Sect. 4.3. From the isotropy of the system it is clear that the propagation speed of uniform wavefronts is independent of their direction of propagation. Thus the virtual propagation speed c' seen in the x direction is given by

$$c' = \frac{\partial x_f}{\partial t} = c \left(1 + \frac{\partial \psi}{\partial t} \right) . \tag{7.3.11}$$

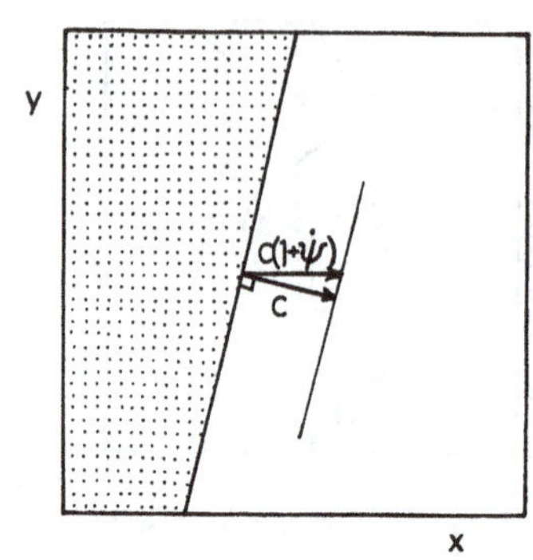

Fig. 7.9. Slightly inclined wavefront, and the relation between its propagating velocities as seen in the x direction and in the direction normal to the wavefront

On the other hand, simple geometrical consideration as depicted in Fig. 7.9 implies

$$c' = c\sqrt{1+c^2\left(\frac{\partial\psi}{\partial y}\right)^2} \simeq c\left[1+\frac{c^2}{2}\left(\frac{\partial\psi}{\partial y}\right)^2\right] \quad (7.3.12)$$

for small $|\partial\psi/\partial y|$. By comparing (7.3.11, 12) with each other, and noting (7.3.9), we have

$$\beta = \frac{c^2}{2}. \quad (7.3.13)$$

This is the very identity we have already proved in Sect. 4.3 on the basis of the general expression for β given by (4.3.26b).

For slowly curved wavefronts in general, we have

$$c\left(1+\frac{\partial\psi}{\partial t}\right) = \tilde{c}\sqrt{1+c^2\left(\frac{\partial\psi}{\partial y}\right)^2} \simeq \tilde{c}+\tilde{c}\beta\left(\frac{\partial\psi}{\partial y}\right)^2 \simeq \tilde{c}+c\beta\left(\frac{\partial\psi}{\partial y}\right)^2, \quad (7.3.14)$$

where $\tilde{c}$ is the local velocity normal to the wavefront, and we have used the fact that $c-\tilde{c}$ is small. Replacing $\partial\psi/\partial t$ by $\alpha\partial^2\psi/\partial y^2+\beta(\partial\psi/\partial y)^2$, we have

$$\tilde{c} = c\left(1+\alpha\frac{\partial^2\psi}{\partial y^2}\right) = c+\alpha\frac{\partial^2 x_f}{\partial y^2} = c+\alpha\varkappa, \quad (7.3.15)$$

where $\varkappa$ is the local front curvature. The last equality gives a coordinate-independent representation of the wavefront dynamics to the lowest-order approximation. Although we started with the assumption that the wavefront is every-

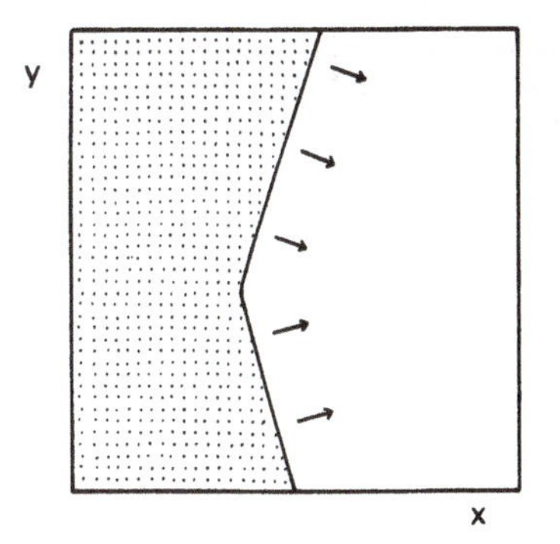

Fig. 7.10. Steadily propagating shock structure as represented by solution (7.3.16)

where almost parallel to the y direction, the final equality in (7.3.15) is free from such a restriction.

When α is positive, (7.3.15) is consistent with the ordinary picture that locally convex fronts tend to be flattened. If a given front is concave, the flattening effect will ultimately be balanced with the sharpening effect (coming from the very fact that the front has a finite propagation velocity), so that formation of a shocklike structure is expected (Fig. 7.10). We already know, in fact, that the nonlinear phase diffusion equation admits a family of shock solutions (though in a different physical context; see Sect. 6.2). In the present notation, the shock solutions (6.2.6) are expressed as

$$\psi(y,t)=\frac{c}{2}(a^2+b^2)t+ay+\frac{2\alpha}{c}\ln\left\{\cosh\left[\frac{bc}{2\alpha}(y+act)\right]\right\}, \tag{7.3.16}$$

where a and b are parameters related to the slope of the front at infinity by

$$\lim_{y\to\pm\infty}\frac{\partial\psi}{\partial y}=a\pm b\,. \tag{7.3.17}$$

7.4 Phase Turbulence

Our principal concern in this section is the behavior of the numerical solutions of the phase turbulence equation (7.2.19) on a finite interval $-\xi/2\leq x\leq\xi/2$, subject to the boundary conditions

$$\frac{\partial\psi}{\partial x}=\frac{\partial^3\psi}{\partial x^3}=0\,,\qquad x=\pm\frac{\xi}{2}. \tag{7.4.1}$$

It is appropriate first to reduce the number of spurious parameters by suitable scaling. To achieve this, we should remember that the phase turbulence equation is valid only for small $|\alpha|$, or for phenomena with a characteristic length of

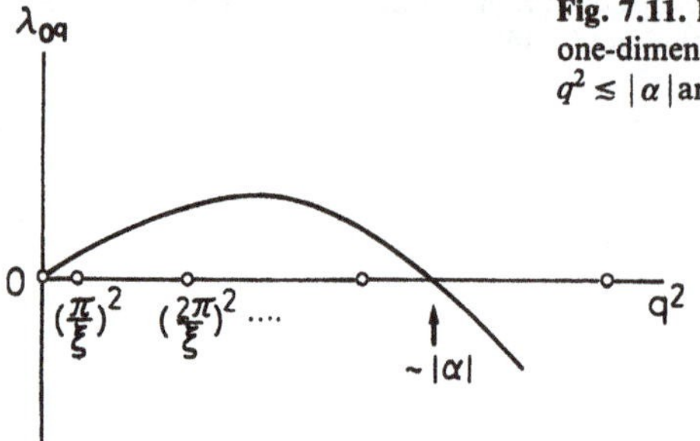

Fig. 7.11. Linear dispersion curve of the phase branch for a one-dimensional system of length ξ. The eigenmodes with $q^2 \lesssim |\alpha|$ are unstable

order $|\alpha|^{-1/2}$. Thus, if ξ is taken to be of order $|\alpha|^{-1/2}$, the number of unstable modes accommodated is of order 1 (possibly zero), which is the case that particularly interests us as far as the onset of chaos is concerned (Fig. 7.11). The system will then behave in much the same way as systems of a few degrees of freedom, and the resulting chaos may possibly be identified with some known type. Instead of assuming that $\xi \sim |\alpha|^{-1/2}$, one could more generally assume

$$\xi = \xi_0 |\alpha|^{-\delta}, \tag{7.4.2}$$

in analogy to the discussion in Sect. 2.3. If $\delta < 1/2$, the uniform solution would be stable, and if $\delta > 1/2$, an infinitely large number of linearly unstable modes would be present as $\alpha \to 0_-$. We now fix δ to be 1/2 and take ξ_0 to be $O(1)$. By transforming ψ, x, and t as

$$\left.\begin{aligned} \psi &\to \beta^{-1}\gamma\left(\frac{\xi_1}{\xi}\right)^2\psi \\ x &\to \frac{\xi}{\xi_1}x \\ t &\to \gamma^{-1}\left(\frac{\xi}{\xi_1}\right)^4 t \end{aligned}\right\}, \tag{7.4.3}$$

where ξ_1 is some constant of $O(1)$, we have

$$\frac{\partial\psi}{\partial t} = -\sigma\frac{\partial^2\psi}{\partial x^2} + \left(\frac{\partial\psi}{\partial x}\right)^2 - \frac{\partial^4\psi}{\partial x^4}, \tag{7.4.4}$$

where

$$\sigma = \gamma^{-1}(\xi_0/\xi_1)^2 . \tag{7.4.5}$$

The system length in the new scale is given by ξ_1. We are left with two parameters σ and ξ_1, but the latter is only spurious; the choice of its value is at our disposal and may be set equal, e.g., to 1. Still, we will retain ξ_1 for computational convenience. We need only remember that the combination $\sigma\xi_1^2$ is the only relevant parameter.

Let ξ_1 be fixed to some value of $O(1)$, and σ increased continuously, so that we can study the routes to chaos. For a given σ, (7.4.4) can be integrated numerically by suitably discretizing x and t. The numerical results for $\psi(x,t)$ obtained are then Fourier-analyzed according to

$$\psi(x,t) = \sum_{\nu} \left[A_\nu(t) \cos\left(\frac{2\nu\pi}{\xi_1}x\right) + B_\nu(t) \sin\left(\frac{(2\nu+1)\pi}{\xi_1}x\right)\right]. \tag{7.4.6}$$

In this way, the phase portrait in the many-dimensional Euclidean space with coordinates $(A_\nu, B_\nu;\ \nu = 1,2,\dots)$ may in principle be obtained. (The uniform mode A_0 does not appear in the evolution equations for the remaining modes, and is unimportant.) Here we note that both the evolution equation and the boundary conditions are invariant under the spatial inversion $x \to -x$. By this transformation any tracjetory $[A_\nu(t), B_\nu(t);\ \nu = 1,2,\dots]$ is changed to $[A_\nu(t), -B_\nu(t);\ \nu = 1,2,\dots]$. If an attractor is invariant under this transformation, the attractor may be called symmetric, and if not, asymmetric.

Since it is impossible to visualize trajectories in too high dimensional spaces, it would be more appropriate to project them onto some subspace E of at most two or three dimensions. One may choose any Fourier modes to construct such a subspace as long as the corresponding Fourier ampitudes are not too small. One appropriate choice would be the A_1-A_2 space. Detailed numerical analysis has never been attempted, and we will content ourselves with the findings described below, showing that the present chaos belongs to the commonest type as far as its onset is concerned. Note that the route to chaos is unique for the present system because there is only one parameter involved.

If the system is uniform, the state point must stay at the origin $(0,0)$ of E, and above a certain positive value of σ this steady state is destabilized. In Fig. 7.12a the steady state is seen to be already unstable, and we have a closed orbit. It was found, however, that a number of bifurcations of different steady states and periodic orbits actually precede this limit cycle behavior, but no chaotic behavior appears as yet. The limit cycle in this figure is symmetric; but as σ is increased, this splits into a pair of asymmetric cycles, one of which is shown in Fig. 7.12b. Then, for each asymmetric orbit, there occurs a sequence of subharmonic bifurcations, which is a sign of approaching chaos (May, 1976; Feigenbaum, 1978). This sequence seems to converge very soon. The trajectory in Fig. 7.12c is considered to be already chaotic as inferred from the corresponding quasi-one-dimensional map taken on the Poincaré section P. This return map is shown in Fig. 7.12d, where X indicates the distance on the A_1-A_2 plane from the origin. Chaos associated with a smooth unimodal one-dimensional map like this is known to be most universal (see, e.g., Collet and Eckmann, 1980), and has been most extensively studied in the past. As the parameter σ is increased further, the position of the left endpoint of our return map is lowered, and it eventually reaches the same level as the right endpoint. This happens precisely when the

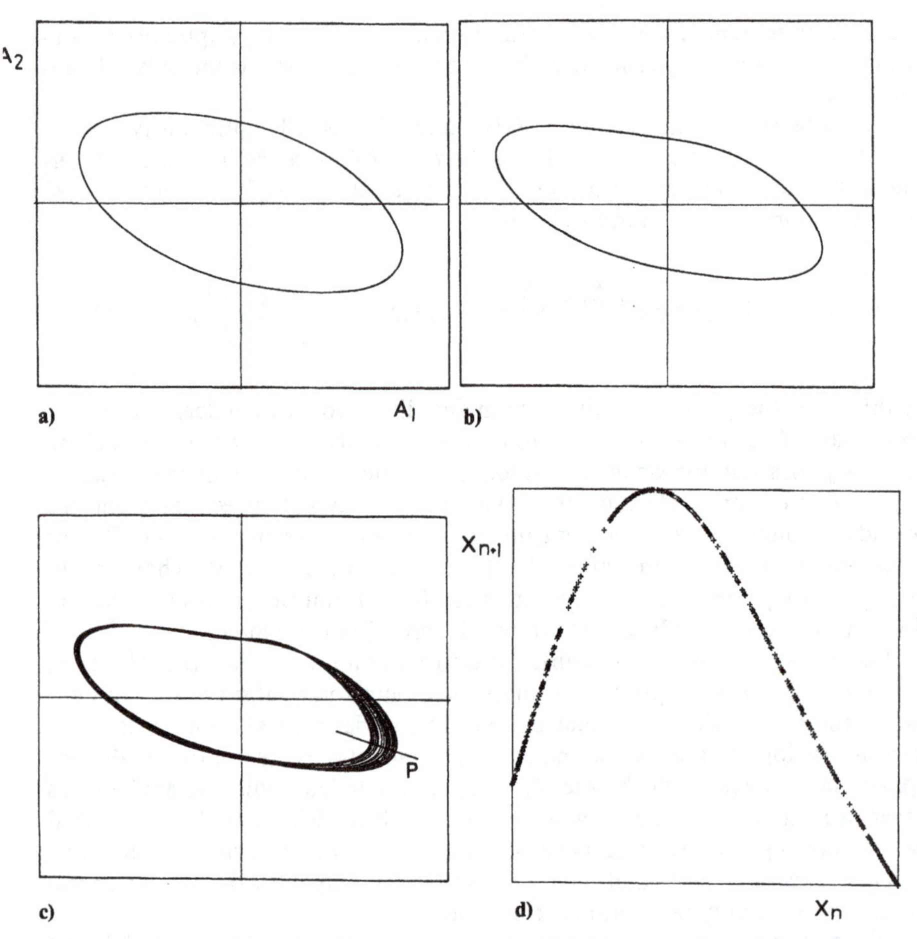

Fig. 7.12 a – d. Some trajectories near the onset of phase turbulence (**a** – **c**), and quasi-one-dimensional map obtained from (**c**) at Poincaré section P (**d**). Parameter values: $\xi_1 = 10.1$ and $\sigma_1 = 2.600$ (**a**), 2.720 (**b**), 2.745 (**c** and **d**)

pair of symmetry-broken chaotic attractors come to join to form a single (symmetric) chaotic attractor.

We would like to know what occurs when σ is made even larger. For intermediate values of σ, there is probably little hope of finding any meaningful description of the dynamics. In the limit of large σ, however, where a large number of effective degrees of freedom would be present, some simple *statistical* description might be possible. In carrying out numerical simulations, however, too large a value of σ is technically inconvenient. Thus, recalling that the only relevant parameter is $\sigma\xi_1^2$, one may alternatively make ξ_1 very large, while σ is fixed to 1. Then we get a chaotic snapshot of ψ (Fig. 7.13), where $t \simeq 200$, for which the initial transient seems to have already been washed out completely. If we collect a sufficiently large number of such snapshots taken at different instants, they would form a statistical ensemble from which we could get various statistical properties of the stationary turbulent state. Space-time correlations of turbulent

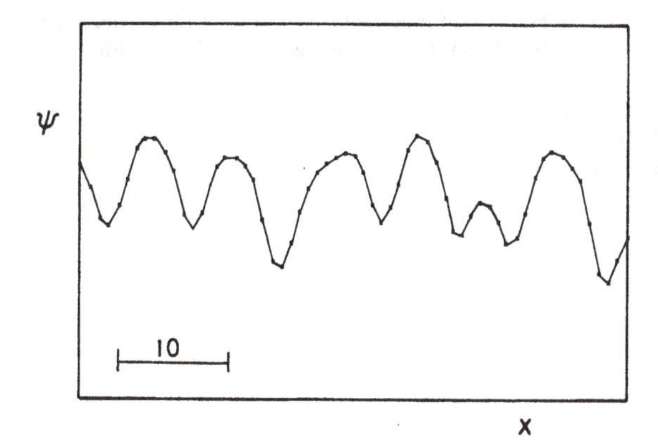

Fig. 7.13. Snapshot of a chaotic phase distribution obtained for the phase turbulence equation (7.4.4) with $\sigma = 1$

phase fluctuations could also be provided from a sufficiently large number of sample pairs formed by $\psi(x,t)$ and $\psi(x',t')$. Extensive numerical calculations of this sort would be valuable, but have never been attempted to date. We only present here a numerical result for the spectrum $S(k)$ of the phase fluctuations in the stationary turbulent state (Yamada and Kuramoto, 1976b), where $S(k)$ is defined by

$$S(k) = \langle |\psi_k|^2 \rangle ,$$

$$\psi_k = \frac{1}{\xi_1} \int_{-\xi_1/2}^{\xi_1/2} \psi(x) e^{ikx} dx . \tag{7.4.7}$$

The average $\langle \ldots \rangle$ is defined by the long-time average

$$\langle A \rangle = \lim_{T\to\infty} \frac{1}{T} \int_0^T A(t) dt . \tag{7.4.8}$$

Figure 7.14 shows the numerically calculated $S(k)$. Two features seem to be worth mentioning. The first is the remarkable peak near $k = 1/\sqrt{2}$, which corresponds to the fluctuation components having the highest linear growth rate. Secondly, the spectrum for smaller wavenumbers seems to obey the law

$$S(k) \propto k^{-2} . \tag{7.4.9}$$

This implies simply that the fluctuation in the phase gradient

$$v(x,t) \equiv \frac{\partial \psi}{\partial x}$$

has a constant intensity, i.e.,

$$\langle |v_k|^2 \rangle \simeq \text{indep. of } k ,$$

so that the "random variable" $v(x,t)$ obeys the central limit theorem just like normal thermal fluctuations. The above property also shows that

$$\langle [\psi(x_0+x) - \psi(x_0)]^2 \rangle \to \infty \quad \text{as} \quad x \to \infty ,$$

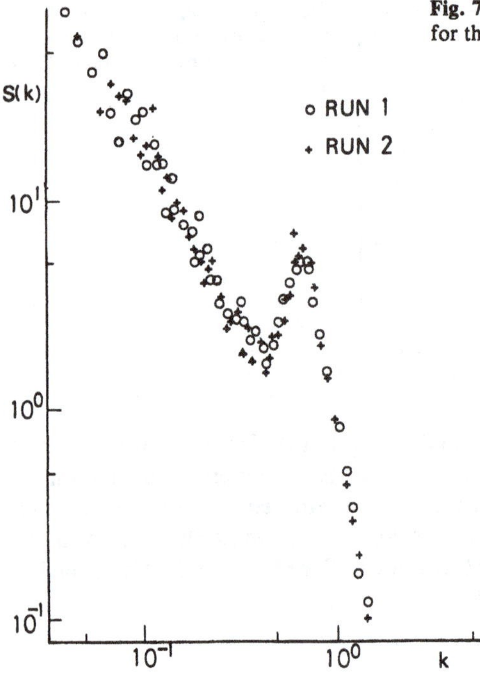

Fig. 7.14. Stationary fluctuation spectrum obtained for the phase turbulence equation (7.4.4) with $\sigma = 1$

implying the loss of long-range coherence in phase. This reminds us of a sufficiently long molecular chain which has no positional long-range order due to the presence of thermal fluctuations, however weak they may be. Finally, we remark that some analytical work (Fujisaka and Yamada, 1977; Yakhot, 1981) and computer-aided studies (Manneville, 1981) exist toward elucidating statistical nature of the phase turbulence equation.

7.5 Amplitude Turbulence

We have seen in Sects. 3.5 and 4.2 that the Ginzburg-Landau equation is appropriate as a model reaction-diffusion system for which the method of phase description is demonstrated. Some coefficients of the expansion of $\partial \psi / \partial t$ where then calculated to give, see (3.5.13a, b and 4.2.27),

$$\alpha = 1 + c_1 c_2, \quad \beta = (c_0 - c_2)(c_2 - c_1), \quad \gamma = -c_1^2(1 + c_2^2)/2.$$

These formulae suggest that the Ginzburg-Landau system exhibits phase turbulence for some suitable range of c_1 and c_2 (and of course for a system length of the order of $|\alpha|^{-1/2}$ or longer). The linear dispersion curves about the uniform oscillations then look like Fig. 7.2a. For stronger instability, as in Fig. 7.2b, the time scales of the unstable phaselike modes would no longer be separated clearly from those of the amplitudelike branch. Even in that case, the number of unstable modes could be made arbitrarily small by adjusting the system length, so that one may then expect a rather simple type of strange attractor to appear. The route to chaos may differ from that for phase turbulence; moreover, there may be many routes, because the system now involves a number of parameters. In the following, we only show a peculiar route to chaos which was found recently (Kuramoto and Koga, 1982). For other routes to chaos, see Moon et al. (1982).

Let us consider the Ginzburg-Landau equation in the form of (2.4.15) on a finite interval $[-\xi/2, \xi/2]$, and require the no-flux boundary conditions

$$\frac{\partial W}{\partial x} = 0, \quad x = \pm\frac{\xi}{2}, \tag{7.5.1}$$

to hold. Numerical calculation was undertaken for values of c_1 and ξ fixed at $c_1 = -2.0$ and $\xi = 3.0$, and c_2 was taken as a bifurcation parameter. It is easy to confirm that the uniform time-periodic solution $W_0(t) = \exp[-\mathrm{i}(c_2 t - \phi_0)]$ (ϕ_0 is an arbitrary phase constant) is linearly stable if $c_2 < c_2^*$ and unstable if $c_2 > c_2^*$, where

$$c_2^* = -\left[2 + (1 + c_1^2)\left(\frac{\pi}{\xi}\right)^2\right] / 2c_1 \simeq 1.87\,. \tag{7.5.2}$$

On the other hand, we found from a computer simulation for the same system that it definitely shows turbulence for sufficiently large c_2. Thus, it was thought that examining bifurcation structures in some intermediate range of c_2 would be interesting.

It should be noted that the Ginzburg-Landau equation subject to the no-flux boundary conditions is invariant under the spatial inversion $x \to -x$, which was also the case for the phase turbulence equation. Although this kind of symmetry property was not very important for the onset of phase turbulence, the same property is crucial to the understanding of the peculiar bifurcation structure in the present case. It is appropriate to make use of the system's symmetry by introducing a complex variable $\tilde{W}(x,t)$ via

$$\tilde{W}(x,t) = W(x,t)\,\mathrm{e}^{-\mathrm{i}\phi(t)} - 1\,, \tag{7.5.3}$$

where $\phi(t)$ stands for the phase of the uniform spatial Fourier component of W. Thus the uniform component of $\tilde{W}$ becomes a real number. One advantage of the representation in terms of $\tilde{W}$ is that the family of uniform oscillations (formed by various ϕ_0 values) falls entirely into the identical state $\tilde{W} = 0$. More generally, the dimension of attractors is lowered by one by working with $\tilde{W}$ instead of W.

Taking account of the boundary conditions (7.5.1), we develop $\tilde{W}$ into Fourier modes as

$$\tilde{W}(x,t) = \sum_{\nu=0}^{\infty}\left[A_\nu(t)\cos\left(\frac{2\nu\pi}{\xi}x\right) + B_\nu(t)\sin\left(\frac{(2\nu+1)\pi}{\xi}x\right)\right]. \tag{7.5.4}$$

Note that A_0 is a real number by definition. As we did for phase turbulence, we may visualize bifurcation structure through the phase portrait in the Hilbert space H formed by A_ν and $B_\nu (\nu = 0, 1, \ldots)$. Remember that the origin $(0, 0, \ldots)$ of H corresponds to the uniform oscillations of various ϕ_0. In order to achieve a better visualization, let us restrict our attention to A_0 and B_0, or equivalently, three real quantities X, Y, and Z defined by

$$X = \mathrm{Re}\{B_0\}, \quad Y = \mathrm{Im}\{B_0\}, \quad Z = -A_0. \tag{7.5.5}$$

The minus sign before A_0 is simply to give a better correspondence to the Lorenz system, i.e., a celebrated three-variable chaos-producing dynamical system (Lorenz, 1963).

We are now interested in the projection of the phase portrait onto the three-dimensional Euclidean space E formed by X, Y, and Z. It is clear that spatial inversion transforms X, Y, and Z as

$$(X, Y, Z) \to (-X, -Y, Z), \tag{7.5.6}$$

which is reminiscent of the Lorenz system and leads to the following properties. Suppose we have an attractor A in space E (correctly speaking, the projection of some attractor in space H onto E). Then $\bar{A}$ is also an attractor, where A and $\bar{A}$ transform to each other via (7.5.6). Of course, A and $\bar{A}$ may happen to be identical; for instance, solution W_0 which appears at the origin $O(0, 0, 0)$ remains invariant under (7.5.6). For c_2 greater than c_2^*, the fixed point at O was seen to become unstable. Just like the Lorenz model, a pair of fixed points P and $\bar{P}$ then bifurcate from O supercritically. Beyond this bifurcation point, however, global analysis would no longer be feasible without the aid of a computer.

The method employed for the analysis of the phase portrait follows that of the preceding section. A computer simulation was carried out directly for the Ginzburg-Landau equation, subject to (7.5.1), with suitable space-time discretization. The numerical data representing $W(x, t)$ were Fourier analyzed to yield $X(t)$, $Y(t)$, and $Z(t)$. The following features were then revealed. With the increase of c_2, the fixed points P and $\bar{P}$ become unstable, and limit cycles L_1 and $\bar{L}_1$ bifurcate from P and $\bar{P}$, respectively. The bifurcation here is of the supercritical type in contrast to the Lorenz system; for the latter system, a subcritical bifurcation occurs at this stage and this leads immediately to chaos. The appearance of a pair of symmetry-broken limit cycles (which as a whole recover symmetry) is reminiscent of the phase turbulence in the situation of Fig. 7.12b. However, what happens thereafter is completely different. As we increase c_2, L_1 and $\bar{L}_1$ come closer to each other and also to the saddle point at O (Fig. 7.15a), and via the formation of a pair of homoclinic orbits at some critical value of c_2, they are joined to form a single closed orbit M_1 (Fig. 7.15b). Now M_1 is invariant under the spatial inversion, i.e., $M_1 = \bar{M}_1$. By further increasing c_2, we found the splitting of M_1 into a pair of closed orbits L_2 and $\bar{L}_2$ (Fig. 7.15c), their recombination into a single closed orbit M_2 (Fig. 7.15d) via the formation of homoclinic orbits, the splitting of M_2 into L_3 and $\bar{L}_3$ (Fig. 7.15e), their recombination into M_3 (Fig. 7.15f), and so on (Fig. 7.15g). Possibly such a process repeats itself an infinite number of times. In general, the L_l are asymmetric (i.e., $L_l \neq \bar{L}_l$) while the M_l are symmetric (i.e., $M_L = \bar{M}_L$). At the moment of each

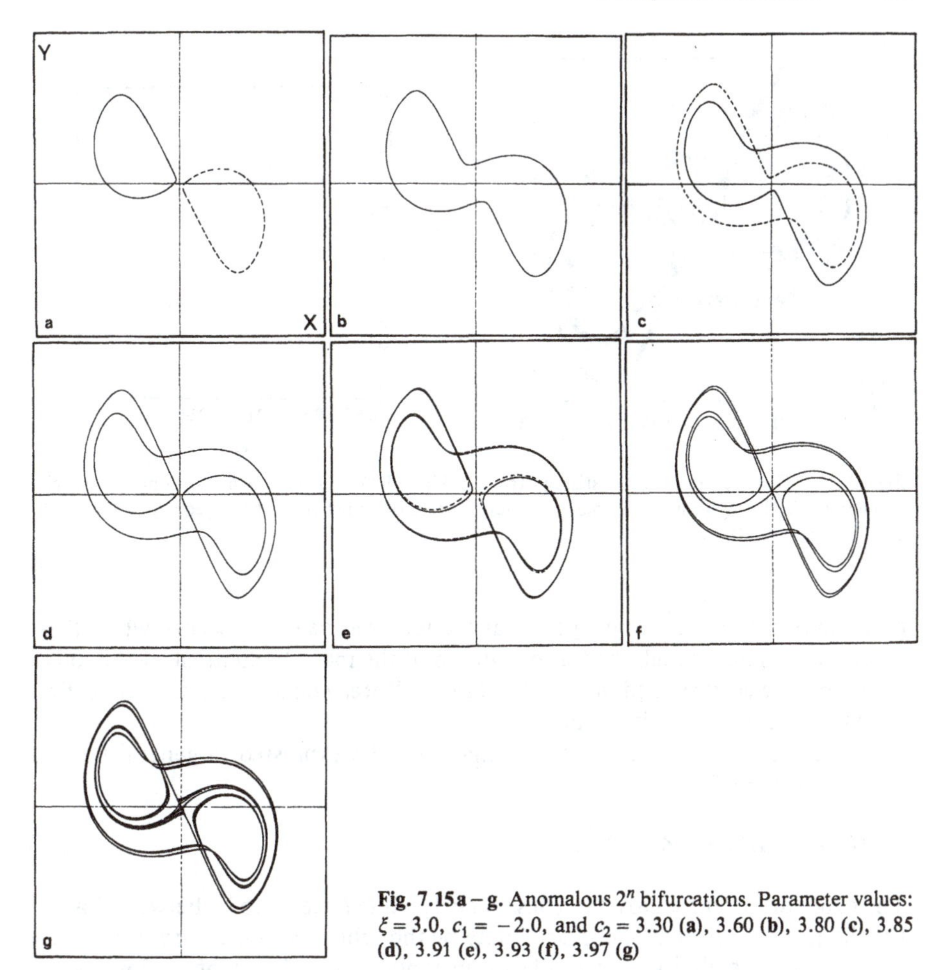

Fig. 7.15a–g. Anomalous 2^n bifurcations. Parameter values: $\xi = 3.0$, $c_1 = -2.0$, and $c_2 = 3.30$ (**a**), 3.60 (**b**), 3.80 (**c**), 3.85 (**d**), 3.91 (**e**), 3.93 (**f**), 3.97 (**g**)

orbital recombination, a pair or homoclinic orbits inevitably appears, and topologically the same thing is expected to happen in the full phase space H. The present type of bifurcation cascade was first discussed by Arneodo et al. (1981). The interval of c_2 between two consecutive bifurcations becomes narrower and narrower, and finally the bifurcations seem to accumulate at some finite value of c_2 which is estimated to be 3.97. Beyond this accumulation point, the attractor (Fig. 7.16) looks much like the classical Lorenz attractor.

Following the method of Lorenz, one may obtain a quasi-one-dimensional map from successive local maxima $Z_n (n = 1, 2, \ldots)$ of $Z(t)$. The result is a unimodal map as shown in Fig. 7.17. Unlike the classical Lorenz chaos, the map seems to have a smooth maximum instead of a cusp structure. The route to chaos, if seen on the map, shows no difference from the usual period-doubling type except that the present map may not have a quadratic maximum; the splitting of M_l into L_{l+1} and $\bar{L}_{l+1}$ appears on the map as the bifurcation of 2^l-point cycles from a 2^{l-1}-point cycle, and the mutual contact of L_{l+1} and $\tilde{L}_{l+1}$ at the saddle

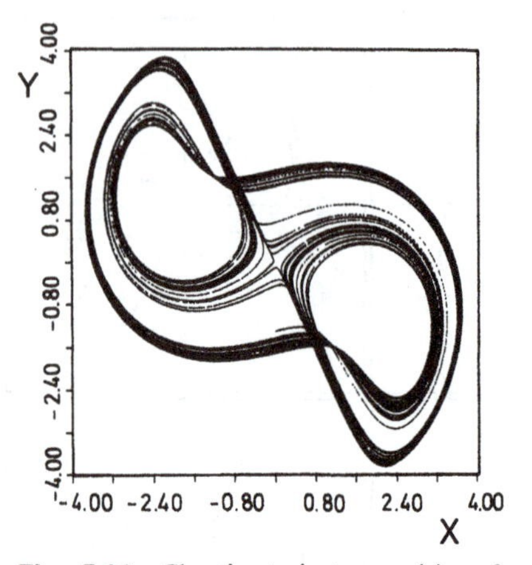

Fig. 7.16. Chaotic trajectory arising from anomalous 2^n bifurcations. Parameter values: $\zeta = 3.0$, $c_1 = -2.00$, $c_2 = 4.00$

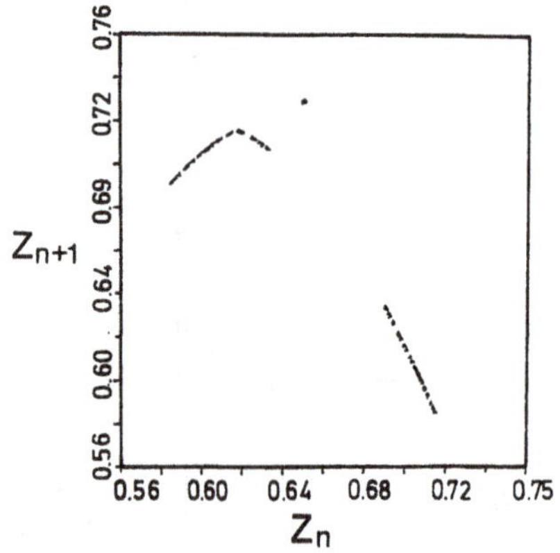

Fig. 7.17. Quasi-one-dimensional Lorenzian map obtained from the trajectory in Fig. 7.16

point makes these 2^l-point cycles superstable because this occurs when these orbits pass right through the flat maximum of the map. In contrast, if the top of the map were cusp shaped, as for the classical Lorenz chaos, then the homoclinic orbits would be infinitely unstable.

Let a unimodal one-dimensional map $f(Z)$ be expressed in general near its maximum point Z^* as

$$f(Z) \simeq f(Z^*) - a|Z - Z^*|^\zeta . \tag{7.5.7}$$

The original Lorenz chaos had ζ less than 1, while ζ seems to lie between 1 and 2 for the present system. The index ζ may be thought of as a measure of how fast the degree of stability (for $\zeta > 1$) or instability (for $\zeta < 1$) of a closed orbit diverges as it transforms itself into a homoclinic orbit. Since the orbits which are almost homoclinic are considered to spend most of their time in the vicinity of the saddle point, the degree of their orbital stability, or ζ, should practically be determined solely from their behavior near the saddle point. There seems to occur an additional simplification from the fact that these orbits, while passing (very slowly) through the vicinity of the saddle point, are expected to fall into a special two-dimensional manifold. This manifold is such that its tangent plane at the saddle point is identical to the eigenplane corresponding to the largest two eigenvalues μ_1 and μ_2 defined for the linearized system about the saddle point. The above argument on ζ essentially follows Yorke and Yorke (1979) who derived the formula

$$\zeta = |\mu_2/\mu_1| , \tag{7.5.8}$$

where $\mu_1 > 0 > \mu_2$ is assumed. The application of this formula certainly gives ζ less than 1 for the classical Lorenz chaos. Analytical calculation of ζ for the present system is also easy. Since the saddle point corresponds to the uniform oscillations, all one has to do is to linearize the Ginzburg-Landau equation about W_0 and find the largest two eigenvalues *except the zero eigenvalue* (i.e., the eigenvalue corresponding to uniform phase fluctuations). The reason for excluding the zero eigenvalue is that the uniform phase fluctuations have entirely been absorbed into the saddle point; this is clear from the definition of $\tilde{W}$. Since the eigenmodes are the Fourier modes, the space E forms itself an eigenspace where the Z axis forms an eigenaxis and the other two eigenaxes lie in the XY plane. For the parameter range of interest to us, space E turns out to be identical to the eigenspace associated with the largest three nonzero eigenvalues. They are given by

$$\mu_z = -2\,,$$

$$\mu_\pm = -\frac{\pi^2}{\xi^2} - 1 \pm \sqrt{1-(2c_1c_2\pi^2/\xi^2)-(c_1^2\pi^4/\xi^4)}\;. \tag{7.5.9}$$

Since $\mu_+ > \mu_z > \mu_-$, we have

$$\zeta = |\mu_z/\mu_+|\;. \tag{7.5.10}$$

This is estimated to give $\zeta \simeq 1.2$ for the parameter values at the accumulation point of the bifurcations. Thus, the value of the Feigenbaum constant δ, in particular, should differ from the standard value 4.669... .

The present study suggests that the probability of encountering smooth one-dimensional maps with non-quadratic maxima should not be ignored as non-generic for real physical systems. The anomalous bifurcation sequence discussed above is a consequence of a system's symmetry with respect to spatial inversion; this kind, or possibly other kinds, of symmetry are commonly present in real physical systems. Experimentally, such a bifurcation sequence could easily be distinguished from the usual subharmonic bifurcations. This is because considerable elongation in period (measured in the continuous time t, and not the step number n) is expected to occur each time a closed orbit is being transformed into a homoclinic orbit.

7.6 Turbulence Caused by Phase Singularities

We discussed rotating spiral waves in Sect. 6.6. on the basis of the Ginzburg-Landau equation, but we were unable to analyze their stability because of the mathematical difficulties involved. Experimentally, spiral waves seem to rotate steadily and rigidly round an almost fixed core. It is reported (Winfree, 1978), however, that more careful observations reveal that the center of the core is not strictly fixed but meanders in a rather irregular way. The analog-computer simulation by Gul'ko and Petrov (1972), and the digital-computer simulation by Rössler and Kahlert (1979) also support the view that such an irregular core motion is common rather than being an exception, especially for excitable reaction-diffusion systems involving "stiff" kinetics. Moreover, there exists the opinion that the core meandering is a form of diffusion-induced chemical turbulence (Rössler and Kahlert, 1979).

It is expected that rotating spiral waves obtained for the Ginzburg-Landau equation become unstable and turbulent if $1+c_1c_2<0$. This is because spiral waves in general behave asymptotically as plane waves far from the core, and under the above condition no plane waves can remain stable. However, we are not much interested in this kind of turbulence in the present section, but we are more interested in the sort of turbulence which would be caused by the very existence of the phase singularity in the core.

Let us restrict our attention, for simplicity, to the case of vanishing c_1, by which the possibility of the first type of turbulence may be eliminated. For infinitely large systems, c_2 is the only parameter involved. Remember that the spiral waves obtained numerically in Chap. 6 were also for $c_1=0$, and that they persisted in being stable for some range of c_2. Here we seek the possibility of their becoming unstable for still larger c_2.

The model equation is given by

$$\frac{\partial W}{\partial t}=W-(1+\mathrm{i}c_2)|W|^2W+\left(\frac{\partial^2}{\partial x^2}+\frac{\partial^2}{\partial y^2}\right)W. \tag{7.6.1}$$

By taking the complex conjugate of this equation, and changing the sign before c_2 at the same time, the equation remains invariant. This says that the only relevant parameter is the absolute value of c_2. As we see below, sufficiently large $|c_2|$ causes turbulence. Since $c_2=\mathrm{Im}\{g\}/\mathrm{Re}\{g\}$, where g is the nonlinear parameter in the original form of the Ginzburg-Landau equation (2.4.10), $|c_2|\to\infty$ as $\mathrm{Re}\{g\}\to 0$ (i.e., as the system approaches the borderline between supercritical and subcritical bifurcations). A number of kinetic models can have parameter values for which $\mathrm{Re}\{g\}=0$, so that such systems should in principle exhibit chemical turbulence of the type discussed below.

At present, no studies exist to show which bifurcations are involved as $|c_2|$ is increased up to the onset of turbulence. We can only show by Fig. 7.18 how a spiral pattern is turbulized starting from a perfectly coherent motion when the parameter value is changed suddenly (Kuramoto and Koga, 1981). Initially $c_2=1.0$ so that a steadily rotating pattern is stable. The initial position of the core has been displaced slightly from the center of symmetry so that axially asymmetric disturbances may be ready to grow whenever the pattern loses stability. We now let c_2 jump to 3.5, and the subsequent temporal development up to $t=12.0$ is shown in the figure. The rotating pattern is apparently unable to adapt smoothly to the new external condition by readjusting its rotation period and wavelength. It becomes increasingly distorted until here and there the contours $\mathrm{Re}\{W\}=0$ and $\mathrm{Im}\{W\}=0$ come into contact with each other; at each moment of such a contact, a new pair of phaseless points at which $W=0$ are produced. Some such phaseless points may soon after be annihilated in pairs, while others may survive for longer periods. Such newly born phaseless points serve themselves as the sources of the subsequent instabilities caused about them. As a result, additional phaseless points will be produced. In this way, we have a cascade process; the turbulent region will spread, and even if the system size is

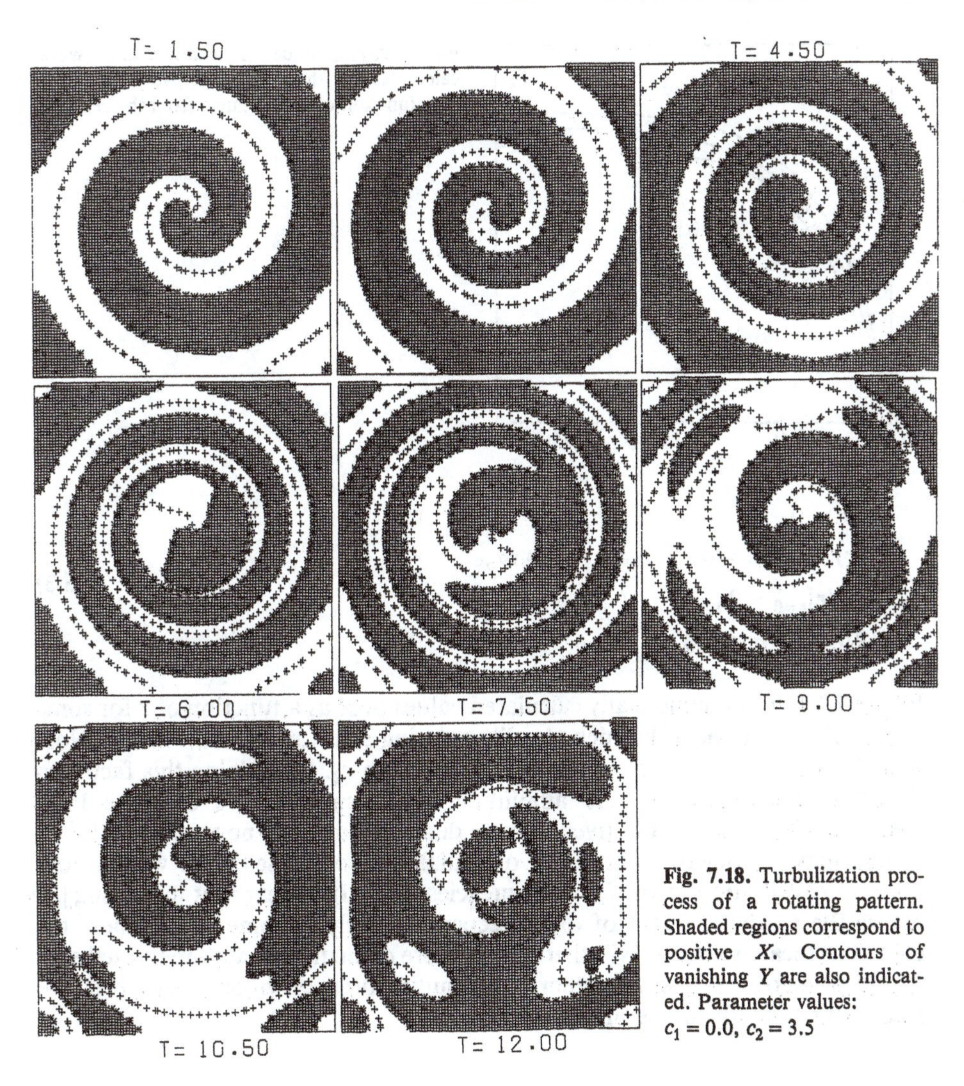

Fig. 7.18. Turbulization process of a rotating pattern. Shaded regions correspond to positive X. Contours of vanishing Y are also indicated. Parameter values: $c_1 = 0.0$, $c_2 = 3.5$

very large, the entire system will eventually become full of such phaseless points. The system could never go back to uniform oscillations however stable the latter state may be; this seems to be particularly true when the initial number of phaseless points is odd, because then there should remain at least one phaseless point which cannot find its counterpart for pair annihilation.

Finally, we give a possible interpretation of how spiral waves are destablized when $|c_2|$ is too large. We remember that a steadily rotating two-dimensional solution of the Ginzburg-Landau equation was obtained in the form

$$W(r,\theta) = R(r)\exp\{i[\Omega t \pm \theta + S(r)]\}, \tag{7.6.2}$$

where R and S have the properties, see (6.6.9a – 10b),

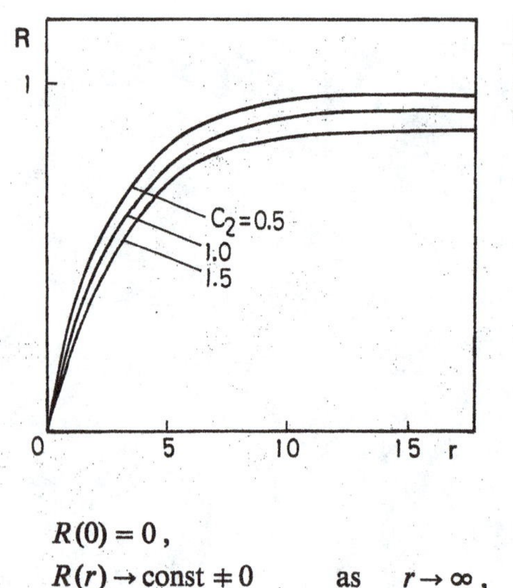

Fig. 7.19. Amplitudes of steadily rotating waves calculated for three different parameter values as a function of the distance from the center of rotation

$$\left.\begin{aligned}
&R(0) = 0\,, && \\
&R(r) \to \text{const} \neq 0 \qquad \text{as} \quad r \to \infty\,, \\
&dS/dr|_{r=0} = 0\,, && \\
&dS/dr \to \text{const} \neq 0 \qquad \text{as} \quad r \to \infty\,.
\end{aligned}\right\} \tag{7.6.3}$$

Figure 7.19 shows numerically calculated values of R as a function of r for some values of c_2, for each of which steadily rotating solutions are still stable. Note that the curves R versus r are rather insensitive to c_2. Combining this fact with (7.2.22) which expresses the amplitude dependence of the effective local frequency $\tilde{\omega}$, we see qualitatively how $\tilde{\omega}$ depends on r. We now realize that the system may be viewed as an array of radially coupled oscillators with a non-uniform distribution of the local frequencies $\tilde{\omega}(r)$. It is clear that increasing $|c_2|$ makes the spatial gradient of $\tilde{\omega}(r)$ steeper especially in some regions near the core. The local oscillators will then find it more difficult to maintain mutual synchronization over the entire system. The resulting breakdown of the synchronization seems to be the cause of turbulence.

Appendix

A. Plane Wave Solutions of the Ginzburg-Landau Equation

The Ginzburg-Landau equation possesses a family of plane wave solutions. They are considered to be a special form of the plane waves whose existence was proved by Kopell and Howard (1973a) for oscillatory reaction-diffusion systems in general. In view of the physical situation where the Ginzburg-Landau equation arises, the plane waves of Kopell and Howard are expected to reduce to this special form as the point of Hopf bifurcation (of the supercritical type) is approached from above. One of the important conclusions to be drawn below is that all the family of plane waves (including uniform oscillation as a special plane wave) can happen to be unstable, which is a property not shared by the $\lambda-\omega$ system with a diagonal diffusion matrix, see Sect. 2.4.

Consider the Ginzburg-Landau equation in the form of (2.4.15) in one space dimension of infinite length:

$$\frac{\partial W}{\partial t} = (1-(1+\mathrm{i}c_2)|W|^2)\,W + (1+\mathrm{i}c_1)\frac{\partial^2 W}{\partial x^2}. \tag{A.1}$$

Equation (A.1) clearly has plane wave solutions

$$W_Q(x,t) = R_Q \exp[\mathrm{i}(Qx-\omega_Q t)]\,, \quad |Q|<1\,, \tag{A.2}$$

where

$$R_Q = \sqrt{1-Q^2}\,, \quad \omega_Q = c_1 Q^2 + (1-Q^2)c_2\,.$$

Note that $|Q|$ may take values between 0 and 1, and that the frequency and amplitude are generally dependent on Q. Of course, such plane waves are highly nonlinear so that their superposition no longer satisfies (A.1).

The stability of W_Q to small perturbations is now investigated. It is convenient to define the deviation u in the form

$$W(x,t) = W_Q(x,t) + u(x,t)\exp[\mathrm{i}(Qx-\omega_Q t)]\,. \tag{A.3}$$

Let (A.1) be linearized in u and its complex conjugate $\bar{u}$, which leads to

$$\frac{\partial u}{\partial t} = \left[-(1+\mathrm{i}c_2)(1-Q^2) - 2(c_1-\mathrm{i})Q\frac{\partial}{\partial x} + (1+\mathrm{i}c_1)\frac{\partial^2}{\partial x^2} \right] u - (1+\mathrm{i}c_2)(1-Q^2)\bar{u} \,. \tag{A.4}$$

Let us work with the Fourier amplitudes $u_q(t)$ of $u(x,t)$ defined by

$$u(x,t) = \int_{-\infty}^{\infty} u_q(t)\,\mathrm{e}^{\mathrm{i}qx} dq \,.$$

Then (A.4) and its complex-conjugate equation yield

$$\frac{d}{dt}\begin{pmatrix} u_q \\ \bar{u}_{-q} \end{pmatrix} = L \begin{pmatrix} u_q \\ \bar{u}_{-q} \end{pmatrix} \tag{A.5}$$

where L is a certain 2×2 matrix. The eigenvalues λ of L are determined from

$$|\lambda - L| = \lambda^2 + (a_1 + \mathrm{i}a_2)\lambda + b_1 + \mathrm{i}b_2 = 0 \,, \tag{A.6}$$

where

$$a_1 = 2(1-Q^2+q^2) \,, \tag{A.7a}$$

$$a_2 = 4c_1 qQ \,, \tag{A.7b}$$

$$b_1 = 2(1+c_1c_2)(1-Q^2)q^2 + (1+c_1^2)q^4 - 4(1+c_1^2)(qQ)^2 \,, \tag{A.7c}$$

$$b_2 = 4(c_1-c_2)(1-Q^2)qQ \,. \tag{A.7d}$$

For given Q and q, the change in the stability property occurs when one of the roots of the quadratic equation (A.6) comes to have a vanishing real part while the real part of the order root remains negative. But we note that $a_1 = -(\mathrm{Re}\{\lambda_1\} + Re\{\lambda_2\}) > 0$, the latter inequality coming from (A.7a) and the property $|Q| < 1$. This means that at least one root of (A.6) has a negative real part for given Q and q. Thus, the critical condition for stability may be found by demanding that some purely imaginary λ satisfies (A.6). This leads to

$$K(Q,q) \equiv b_2^2 - a_1 a_2 b_2 - a_1^2 b_1 = 0 \,. \tag{A.8}$$

As is easily checked, negative $K(Q,q)$ implies stability, and positve $K(Q,q)$ instability. The stability of W_Q discussed above concerns only fluctuations of a given wavenumber q. Thus the complete stability of W_Q requires that

$$\mathrm{Max}_q K(Q,q) \le 0 \,. \tag{A.9}$$

Note also the identity

$$K(Q,0)=0\,, \tag{A.10}$$

which reflects the obvious fact that the plane waves are neutrally stable under spatial translation.

The stability criterion is now examined more closely for a few special circumstances. First, suppose $|Q|\simeq 1$, which means that the amplitudes of the plane waves are small. Then,

$$K(Q,q)\simeq K(1,q)=-4(1+c_1^2)(q^2-4)q^6\,. \tag{A.11}$$

Since this quantity is positive for sufficiently small nonzero $|q|$, the plane waves are always unstable. The origin of the instability of small-amplitude plane waves is also clear from the fact that the zero-amplitude plane wave state is nothing but the unstable steady state from which the time-periodic solution has bifurcated.

Consider the other extreme, i.e., $Q=0$, the corresponding "plane wave" being uniform oscillation. Then,

$$K(0,q)=-8(1+q^2)^2q^2\varkappa(q)\,,\quad \text{where} \tag{A.12}$$

$$\varkappa(q)=1+c_1c_2+\frac{1+c_1^2}{2}q^2\,. \tag{A.13}$$

The quantity α defined by

$$\alpha=1+c_1c_2 \tag{A.14}$$

is now seen to be an important parameter to the stability of the uniform oscillation; if $\alpha>0$, $K(0,q)$ satisfies (A.9), implying stability, whereas if $\alpha<0$, $K(0,q)$ cannot satisfy (A.9) for sufficiently small q, implying instability. α can happen to be negative in some concrete models (Appendix B). It is concluded, in particular, that the stability of the uniform limit cycle oscillation is not guaranteed for infinitely large system size even if this oscillation is a small-amplitude solution which has just bifurcated supercritically.

The eigenvalue spectrum of the fluctuations about the uniform oscillation is obtained from

$$\lambda^2+2(1+q^2)\lambda+2(1+c_1c_2)q^2+(1+c_1^2)q^4=0\,. \tag{A.15}$$

The roots of this equation are given by

$$\begin{aligned}\lambda_{\mathrm{p}}&=-(1+q^2)\left[1-\sqrt{1-2\left(\frac{q}{1+q^2}\right)^2\varkappa(q)}\,\right],\\ \lambda_{\mathrm{a}}&=-(1+q^2)\left[1+\sqrt{1-2\left(\frac{q}{1+q^2}\right)^2\varkappa(q)}\,\right].\end{aligned} \tag{A.16}$$

Note that $\lambda_p \to 0$ and $\lambda_a \to -2$ as $q \to 0$. The spectrum of λ_p may be called the phaselike branch, and that of λ_a the amplitudelike branch, since they are associated, respectively, with phase fluctuations and amplitude fluctuations in the limit of vanishing q. We may say that the instability of the uniform oscillation is related to the unstable growth of some phaselike fluctuations.

Finally, we consider the critical condition for general Q but only with respect to perturbations with small $|q|$. The stability condition $K(Q,q)<0$ $(q \neq 0)$ then reduces to

$$(1+c_1c_2)(1-Q^2)-2(1+c_2^2)Q^2>0\,. \tag{A.17}$$

Note that there exist no stable plane waves if $\alpha<0$. On the contrary, if $\alpha>0$, there exists a critical value Q_c of Q given by

$$Q_c=\left(\frac{\alpha}{\alpha+2(1+c_2^2)}\right)^{1/2}, \tag{A.18}$$

such that any plane wave with $|Q|$ greater than Q_c is unstable. For $|Q|$ smaller than Q_c, the plane waves are stable at least to long-wavelength fluctuations. In any case, the uniform oscillation is the last "plane wave" to be destabilized as α is descreased.

B. The Hopf Bifurcation for the Brusselator

The hypothetical chemical reaction model called the Brusselator which was proposed by the Brussels school (Glansdorff and Prigogine, 1971) serves as a particularly convenient model for illustrating various results obtained in the text. In this appendix, the method developed in Chap. 2 will be illustrated for this model. Specifically, we try to reduce the Brusselator, including diffusion, to the Ginzburg-Landau equation, thereby calculating coefficients λ_1, d, and g explicitly and discussing their physical implications.

The chemical basis of the Brusselator is not of our concern here; we will only be concerned with its nature as a dynamical system. It is a two-component system whose simplest version in the presence of diffusion takes the form

$$\begin{aligned}\frac{\partial X}{\partial t}&=A-(B+1)X+X^2Y+D_X\nabla^2X\,,\\ \frac{\partial Y}{\partial t}&=BX-X^2Y+D_Y\nabla^2Y\,,\end{aligned} \tag{B.1}$$

where A, B, D_X, and D_Y are non-negative constants. In what follows, the system size is assumed to be infinite. There exists a unique uniform steady state (X_0, Y_0) given by

$$(X_0, Y_0) = (A, B/A). \tag{B.2}$$

Defining fluctuations ξ and η by $(X, Y) = (X_0 + \xi, Y_0 + \eta)$, we have

$$\frac{\partial \xi}{\partial t} = (B-1)\xi + A^2\eta + D_X\nabla^2\xi + f(\xi,\eta),$$
$$\frac{\partial \eta}{\partial t} = -B\xi - A^2\eta + D_Y\nabla^2\eta - f(\xi,\eta), \tag{B.3}$$

where

$$f(\xi,\eta) = \frac{B}{A}\xi^2 + 2A\xi\eta + \xi^2\eta. \tag{B.4}$$

Suppose A is kept constant and B is taken to be a control parameter. The stability of (X_0, Y_0) may be analyzed as follows: linearizing (B.3) in ξ and η, and assuming that $\xi, \eta \propto \exp(\mathrm{i}\boldsymbol{q}\boldsymbol{r} + \lambda t)$, we have

$$\lambda^2 + \alpha(q)\lambda + \beta(q) = 0, \tag{B.5}$$

where $q \equiv |\boldsymbol{q}|$ and

$$\alpha(q) = 1 + A^2 - B + (D_X + D_Y)q^2, \tag{B.6a}$$

$$\beta(q) = A^2 + [A^2 D_X + (1-B)D_Y]q^2 + D_X D_Y q^4. \tag{B.6b}$$

The steady state (X_0, Y_0) is linearly stable if and only if both $\alpha(q)$ and $\beta(q)$ are non-negative for all q. Clearly, this stability condition can be violated in either of the following two ways:

1) $\alpha(q)$ vanishes for some q, but otherwise $\alpha(q)$ and $\beta(q)$ remain positive for all q.
2) $\beta(q)$ vanishes for some q, but otherwise $\alpha(q)$ and $\beta(q)$ remain positive for all q.

For Type 1 instability, the critical value of q is zero. Then the critical B value is given by

$$B_c = 1 + A^2. \tag{B.7}$$

Clearly, $B < B_c$ implies stability, and $B > B_c$ instability. For Type 2 instability, the critical B value (denoted as B_c') and critical wavenumber q_c are obtained from the conditions $\beta(q_c) = 0$ and $d\beta(q_c)/dq_c = 0$. We have

$$B_c' = (1 + A\sqrt{D_X/D_Y})^2, \tag{B.8}$$

$$q_c^2 = A/\sqrt{D_X D_Y}. \tag{B.9}$$

The possibility of Type 2 instability for general reaction-diffusion systems was first pointed out mathematically by Turing (1952). Let us now assume $B_c < B_c'$ or, equivalently,

$$\gamma \equiv \sqrt{D_X/D_Y} > A^{-1}(\sqrt{1+A^2}-1)\,, \tag{B.10}$$

to eliminate the possibility of Type 2 instability occurring first as B is increased. In particular, the condition $D_X > D_Y$ is sufficient in this inequality because the right-hand side of (B.10) is definitely smaller than 1.

Since we have a pair of complex-conjugate eigenvalues which become purely imaginary, $\pm iA$, at $B = B_c$, the method of Chap. 2 is applicable near the critical point. Let μ be defined by

$$\mu = (B - B_c)/B_c\,. \tag{B.11}$$

Some quantities necessary for calculating λ_1, d, and g are the following:

$$L_0 = \begin{pmatrix} A^2 & A^2 \\ -(1+A^2) & -A^2 \end{pmatrix}, \tag{B.12a}$$

$$D = \begin{pmatrix} D_X & 0 \\ 0 & D_Y \end{pmatrix}, \tag{B.12b}$$

$$L_1 = (1+A^2)\begin{pmatrix} 1 & 0 \\ -1 & 0 \end{pmatrix}, \tag{B.12c}$$

$$u_0 = \begin{pmatrix} 1 \\ -1+iA^{-1} \end{pmatrix}, \tag{B.12d}$$

$$u_0^* = \tfrac{1}{2}(1-iA, -iA)\,. \tag{B.12e}$$

The application of (2.2.5b) and (2.4.11) immediately leads to

$$\lambda_1 = \frac{1+A^2}{2}, \tag{B.13}$$

$$d = \tfrac{1}{2}[D_X + D_Y - iA(D_X - D_Y)]\,. \tag{B.14}$$

Note that λ_1 is purely real, which is peculiar to the present model. The calculation of g is a little more cumbersome. Note first that the nonlinear terms $f(\xi, \eta)$ in (B.3) still contain the parameter B. This is replaced by B_c to give M_0 and N_0. Specifically, the ξ and η components of the vectors such as $M_0\boldsymbol{ab}$ and $N_0\boldsymbol{abc}$ are given by

$$(M_0 ab)_\xi = -(M_0 ab)_\eta = \frac{1+A^2}{A} a_\xi b_\xi + A(a_\xi b_\eta + a_\eta b_\xi),$$

$$\begin{aligned}(N_0 abc)_\xi &= -(N_0 abc)_\eta \\ &= \tfrac{1}{3}(a_\xi b_\xi c_\eta + a_\xi b_\eta c_\xi + a_\eta b_\xi c_\xi).\end{aligned} \tag{B.15}$$

Thus, we get via (2.2.17b),

$$\begin{aligned} V_+ &= \bar{V}_- = \frac{(1+\mathrm{i}A)^3}{3A^3}\begin{pmatrix} -2\mathrm{i}A \\ 1+2\mathrm{i}A \end{pmatrix}, \\ V_0 &= \frac{2(A^2-1)}{A^3}\begin{pmatrix} 0 \\ 1 \end{pmatrix}. \end{aligned} \tag{B.16}$$

The substitution of (B.12d, e, 15, 16) into (2.2.20) gives

$$g = \frac{1}{2}\left(\frac{2+A^2}{A^2} + \mathrm{i}\frac{4-7A^2+4A^4}{3A^3}\right). \tag{B.17}$$

Since $\mathrm{Re}\{g\} > 0$, we always have a supercritical Hopf bifurcation.

If the Ginzburg-Landau equation corresponding to the present model is expressed in the reduced form (2.4.15), then the parameters involved are given by

$$c_1 = -A\frac{D_X - D_Y}{D_X + D_Y} = -A\frac{\gamma^2-1}{\gamma^2+1}, \tag{B.18a}$$

$$c_2 = \frac{4-7A^2+4A^4}{3A(2+A^2)}. \tag{B.18b}$$

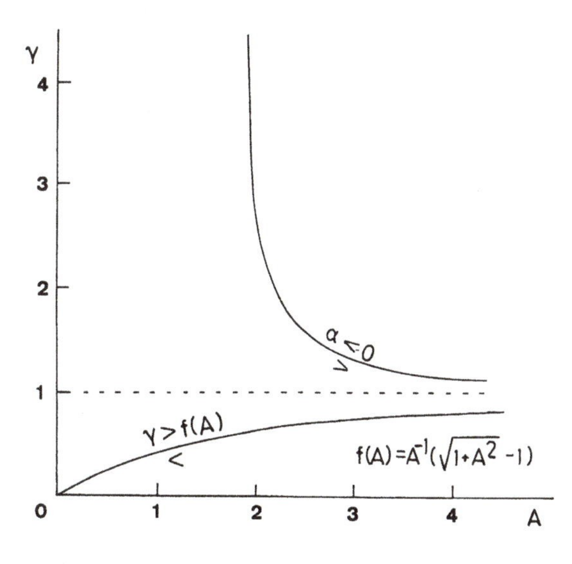

Fig. B.1. In the parameter region $\gamma > f(A)$ the onset of spatially uniform oscillation precedes the onset of spatially non-uniform structure. If, in addition, α is negative, the resulting oscillation is unstable with respect to long-scale phase fluctuations. The figure shows the existence of such a parameter region for the Brusselator

The important parameter α defined in (A.14) is expressed as

$$\alpha = 1 - \frac{\gamma^2 - 1}{3(\gamma^2 + 1)} \frac{4 - 7A^2 + 4A^4}{2 + A^2}. \tag{B.19}$$

Let us examine whether the condition $\alpha < 0$ (i.e., the instability condition for the uniform oscillation due to diffusion) is possible for the Brusselator. Note that we are always subject to condition (B.10). As Fig. B.1 shows, there certainly exists a parameter region where the two conditions, i.e., $\alpha < 0$ and (B.10), are satisfied simultaneously.

Finally, we note that $|c_1|$, $c_2 \to \infty$ as $A \to \infty$. According to the discussion in Sect. 2.4, this means that the Brusselator with diffusion behaves like a nonlinear Schrödinger equation slightly above the Hopf bifurcation point, provided A is sufficiently large.

References

Agladze, K. I., Krinsky, V. I. (1982): Multi-armed vortices in an excitable chemical medium. Nature **296**, 424

Aizawa, Y. (1976): Synergetic approach to the phenomena of mode-locking in nonlinear systems. Prog. Theor. Phys. **56**, 703

Allessie, M. A., Bonke, F. I. M., Shopman, F. J. G. (1977): Circus movement in rabbit atrial muscle as a mechanism of tachycardia. III. The "leading circle" concept: A new model of circus movement in cardiac tissue without the involvement of an anatomical obstacle. Circ. Res. **41**, 9

Arneodo, A., Coullet, P., Tresser, C. (1981): A possible new mechanism for the onset of turbulence. Phys. Lett. **81A**, 197

Auchmuty, J. F. G., Nicolis, G. (1975): Bifurcation analysis of nonlinear reaction-diffusion equations I. Evolution equations and the steady state solutions. Bull. Math. Biol. **37**, 323

Auchmuty, J. F. G., Nicolis, G. (1976): Bifurcation analysis of nonlinear reaction-diffusion equations III. Chemical oscillations. Bull. Math. Biol. **38**, 325

Bogoliubov, N. N., Mitropolskii, Y. A. (1961): *Asymptotic Methods in the Theory of Nonlinear Oscillations* (Gordon and Breach, New York) [English transl.]

Bünning, E. (1973): *The Phyiological Clock*, 3rd ed. (Springer, New York)

Burgers, J. M. (1974): *The Nonlinear Diffusion Equation – Asymptotic Solutions and Statistical Physics* (Reidel, Dordrecht)

Cesari, L. (1971): *Asymptotic Behavior and Stability Problems in Ordinary Differential Equations* (Springer, New York)

Coddington, E. A., Levinson, N. (1955): *Differential Equations* (McGraw-Hill, New York)

Cohen, D. S., Neu, J. C., Rosales, R. R. (1978): Rotating spiral wave solutions of reaction-diffusion equations. SIAM J. Appl. Math. **35**, 536

Collet, P., Eckmann, J. P. (1980): *Iterated Maps of the Interval as Dynamical Systems* (Birkhauser, Boston)

DiPrima, R. C., Swinney, H. L. (1981): Instabilities and transition in flow between concentric rotating cylinders. In *Hydrodynamic Instabilities and the Transition to Turbulence*, ed. by H. L. Swinney, J. P. Gollub, Topics Appl. Phys., Vol. 45 (Springer, Berlin, Heidelberg, New York) p. 139

Ermentrout, G. B. (1982): Asymptotic behavior of stationary homogeneous neural nets. In *Competition and Cooperation in Neural Nets*, ed. by S. Amari, M. A. Arbib, Lecture Notes in Biomath. Vol. 45 (Springer, Berlin, Heidelberg, New York) p. 57

Erneux, T., Herschkowitz-Kaufman, M. (1977): Rotating waves as asymptotic solutions of a model chemical reaction. J. Chem. Phys. **66**, 248

Feigenbaum, M. J. (1978): Quantitative universality for a class of nonlinear transformations. J. Stat. Phys. **19**, 25

Fenstermacher, P. R., Swinney, H. L., Gollub, J. P. (1979): Dynamical instabilities and the transition to chaotic Taylor vortex flow. J. Fluid Mech. **94**, 103

Field, R. J., Körös, E., Noyes, R. M. (1972): Oscillations in chemical systems. II. Thorough analysis of temporal oscillations in the bromate-cerium-malonic acid system. J. Am. Chem. Soc. **94**, 8649

Fife, P. C. (1976a): Singular perturbation and wave front techniques in reaction-diffusion problems. SIAM-AMS Proc. **10**, 23

Fife, P. C. (1976b): Pattern formation in reacting and diffusing systems. J. Chem. Phys. **14**, 554

Fife, P. C. (1979a): *Mathematical Aspects of Reacting and Diffusing Systems*, Lecture Notes Biomath., Vol. 28 (Springer, Berlin, Heidelberg, New York)

Fife, P. C. (1979b): Wave-fronts and target patterns. In *Applications of Nonlinear Analysis in the Physical Sciences*, ed. by H. Amann, N. Bazley, K. Kirchgassner (Pitman, London) p. 206

FitzHugh, R. (1961): Impulses and physiological states in theoretical models of nerve membrane. Biophys. J. **1**, 445
FitzHugh, R. (1969): Mathematical models of excitation and propagation in nerve. In *Biological Engineering*, ed. by H. P. Schwan (McGraw-Hill, New York) p. 1
Fujii, H., Sawada, Y. (1978): Phase difference locking of coupled oscillating chemical systems. J. Chem. Phys. **69**, 3830
Fujisaka, H., Yamada, T. (1977): Theoretical study of chemical turbulence. Prog. Theor. Phys. **57**, 734
Gerisch, G. (1968): Cell aggregation and differentiation in *Dictyostelium*. In *Current Topics in Developmental Biology* **3**, ed. by A. Moscona, A. Monroy (Academic, New York) p. 157
Gibbon, J. D., McGuiness, M. J. (1981): Amplitude equations at the critical points of unstable dispersive physical systems. Proc. Roy. Soc. London **A377**, 185
Gierer, A., Meinhardt, H. (1972): A theory of biological pattern formation. Kybernetik **12**, 30
Giglio, M., Musazzi, S., Perini, U. (1981): Transition to chaotic behavior via a reproducible sequence of period-doubling bifurcations. Phys. Rev. Lett. **47**, 243
Glansdorff, P., Prigogine, I. (1971): *Thermodynamic Theory of Structure, Stability, and Fluctuations* (Wiley, London)
Gollub, J. P., Benson, S. V. (1980): Many routes to turbulent convection. J. Fluid Mech. **100**, 449
Graham, R., Haken, H. (1968): Quantum theory of light propagation in a fluctuating laser-active medium. Z. Phys. **213**, 420
Graham, R., Haken, H. (1970): Laserlight – first example of a second-order phase transition far away from thermal equilibrium. Z. Phys. **237**, 31
Greenberg, J. M. (1976): Periodic solutions to reaction-diffusion equations. SIAM J. Appl. Math. **30**, 199
Greenberg, J. M. (1978): Axi-symmetric, time-periodic solutions of reaction-diffusion equations. SIAM J. Appl. Math. **34**, 391
Greenberg, J. M. (1980): Spiral waves for $\lambda-\omega$ systems. SIAM J. Appl. Math. **39**, 301
Guckenheimer, J. (1975): Isochrons and phaseless sets. J. Math. Biol. **1**, 259
Guckenheimer, J. (1981): On codimension two bifurcations. In *Dynamical Systems and Turbulence, Warwick 1980*, ed. by D. A. Rand, L. S. Young, Lecture Notes in Math. (Springer, Berlin, Heidelberg, New York) p. 99
Gul'ko, F. B., Petrov, A. A. (1972): Mechanism of the formation of closed pathways of conduction in excitable media. Biofizika **17**, 71
Haken, H., Sauermann, H. (1963): Frequency shifts of laser modes in solid state and gaseous systems. Z. Phys. **176**, 47
Haken, H. (1975a): Generalized Ginzburg-Landau equations for phase transitionlike phenomena in lasers, nonlinear optics, hydrodynamics and chemical reactions. Z. Phys. **B21**, 105
Haken, H. (1975b): Cooperative phenomena in systems far from thermal equilibrium and in nonphysical systems. Rev. Mod. Phys. **47**, 67
Haken, H. (1983a): *Synergetics – An Introduction: Nonequilibrium Phase Transitions and Self-Organization in Physics, Chemistry and Biology*, 3rd ed. (Springer, Berlin, Heidelberg, New York)
Haken, H. (1983b): *Advanced Synergetics – Instability Hierarchies of Self-Organizing Systems and Devices*, Springer Ser. Syn., Vol. 20 (Springer, Berlin, Heidelberg, New York)
Haken, H., Wunderlin, A. (1982): Slaving principle for stochastic differential equations with additive and multiplicative noise and for discrete noisy maps. Z. Phys. **47**, 179
Hastings, S. P. (1976): Periodic plane waves for the Oregonator. Stud. Appl. Math. **55**, 293
Herschkowitz-Kaufman, M. (1975): Bifurcation analysis of nonlinear reaction-diffusion equations II. Steady state solutions and comparison with numerical simulations. Bull. Math. Biol. **37**, 589
Howard, L. N., Kopell, N. (1977): Slowly varying waves and shock structures in reaction-diffusion equations. Stud. Appl. Math. **56**, 95
Hudson, J. L., Hart, M., Marinko, D. (1979): An experimental study of multiple peak periodic and nonperiodic oscillations in the Belousov-Zhabotinskii reaction. J. Chem. Phys. **71**, 1601
Ivanitsky, G. R., Krinsky, V. I., Zaikin, A. N., Zhabotinsky, A. M. (1981): Autowave processes and their role in disturbing the stability of distributed excitable systems. In *Soviet Scientific Reviews*, Section D, Biological Reviews 2, ed. by V. P. Skulachev (Soviet Scientific Reviews) p. 279
Joseph, D. D., Sattinger, D. H. (1972): Bifurcating time periodic solutions and their stability. Arch. Rational. Mech. Anal. **45**, 79

Karfunkel, H. R., Seelig, F. F. (1975): Excitable chemical reaction systems I. Definition of excitability and simulation of model systems. J. Math. Biol. **2**, 123

Karfunkel, H. R., Kahlert, C. (1977): Excitable chemical reaction systems II. Several pulses on the ring fiber. J. Math. Biol. **4**, 183

Kidachi, H. (1980): On mode interactions in reaction-diffusion equation with nearly degenerate bifurcations. Prog. Theor. Phys. **63**, 1152

Koga, S. (1982): Rotating spiral waves in reaction-diffusion systems – Phase singularities and multi-armed waves. Prog. Theor. Phys. **67**, 164

Kopell, N., Howard, L. N. (1973a): Plane wave solutions to reaction-diffusion equation. Stud. Appl. Math. **52**, 291

Kopell, N., Howard, L. N. (1973b): Horizontal bands in the Belousov reaction. Science **180**, 1171

Kuramoto, Y., Tsuzuki, T. (1974): Reductive perturbation approach to chemical instabilities. Prog. Theor. Phys. **52**, 1399

Kuramoto, Y. (1975): Self-entrainment of a population of coupled nonlinear oscillators. In *International Symposium on Mathematical Problems in Theoretical Physics*, ed. by H. Araki, Lecture Notes Phys., Vol. 39 (Springer, New York) p. 420

Kuramoto, Y., Tsuzuki, T. (1976): Persistent propagation of concentration waves in dissipative media far from thermal equilibrium. Prog. Theor. Phys. **55**, 356

Kuramoto, Y., Yamada, T. (1976a): Turbulent state in chemical reactions. Prog. Theor. Phys. **55**, 679

Kuramoto, Y., Yamada, T. (1976b): Pattern formation in oscillatory chemical reactions. Prog. Theor. Phys. **56**, 724

Kuramoto, Y. (1978): Diffusion-induced chaos in reaction systems. Prog. Theor. Phys. Suppl. **64**, 346

Kuramoto, Y. (1980a): Instability and turbulence of wavefronts in reaction-diffusion systems. Prog. Theor. Phys. **63**, 1885

Kuramoto, Y. (1980b): Diffusion-induced chemical turbulence. In *Dynamics of Synergetic Systems*, ed. by H. Haken, Springer Ser. Syn., Vol. 6 (Springer, Berlin, Heidelberg, New York) p. 134

Kuramoto, Y. (1981): Rhythms and turbulence in populations of chemical oscillators. Physica **106A**, 128

Kuramoto, Y., Koga, S. (1981): Turbulized rotating chemical waves. Prog. Theor. Phys. **66**, 1081

Kuramoto, Y., Koga, S. (1982): Anomalous period-doubling bifurcations leading to chemical turbulence. Phys. Lett. **92A**, 1

Landau, L. D. (1944): On the problem of turbulence. C. R. Dokl. Acad. Sci. URSS **44**, 311

Libchaber, A., Maurer, J. (1980): Une expérience de Rayleigh-Bénard de géométrie réduite: Multiplication, accrochage et démultiplication de fréquences. J. Phys. (Paris) Colloq **41**, C3, 51

Lin, J., Kahn, P. B. (1982): Phase and amplitude instability in delay-diffusion population models. J. Math. Biol. **13**, 383

Lorenz, E. N. (1963): Deterministic nonperiodic flow. J. Atmos. Sci. **20**, 130

Manneville, P. (1981): Statistical properties of chaotic solutions of a unidimensional model for phase turbulence. Phys. Lett. **84A**, 129

Marek, M., Stuchl, I. (1975): Synchronization in two interacting oscillating systems. Biophys. Chem. **3**, 241

Marsden, J. E., McCracken, M. (1976): *The Hopf Bifurcation and Its Applications*, Appl. Math. Sci., Vol. 19 (Springer, New York)

May, R. M. (1976): Simple mathematical models with very complicated dynamics. Nature **261**, 459

McKean, H. P. (1970): Nagumo's equation. Adv. Math. **4**, 209

Michelson, D. M., Sivashinsky, G. I. (1977): Nonlinear analysis of hydrodynamic instability in laminar flames – II. Numerical experiments. Acta Astronautica **4**, 1207

Moon, H. T., Huerre, P., Redekopp, L. G. (1982): Three-frequency motion and chaos in the Ginzburg-Landau equation. Phys. Rev. Lett. **49**, 458

Murray, J. D. (1976): On travelling wave solutions in a model for the Belousov-Zhabotinsky reaction. J. Theor. Biol. **56**, 329

Murray, J. D. (1977): *Lectures on Nonlinear-Differential-Equation Models in Biology* (Clarendon, Oxford)

Musha, T., Kosugi, Y., Matsumoto, G. (1981): Modulation of the time relation of action potential impulses propagating along an axon. IEEE Trans. Biomed. Eng. **28**, 616

Nagashima, H. (1980): Chaotic states in the Belousov-Zhabotinsky reaction. J. Phys. Soc. Japan **49**, 2427

Nagumo, J., Arimoto, S., Yoshizawa, S. (1962): An active pulse transmission line simulating nerve axon. Proc. IRE **50**, 2061

Nayfeh, A. H. (1973): *Perturbation Methods* (Wiley, New York)

Neu, J. C. (1979a): Chemical waves and the diffusive coupling of limit cycle oscillators. SIAM J. Appl. Math. **36**, 509

Neu, J. C. (1979b): Coupled chemical oscillators. SIAM J. Appl. Math. **37**, 307

Neu J. C. (1980): Large populations of coupled chemical oscillators. SIAM J. Appl. Math. **38**, 305

Newell, A. C., Whitehead, J. A. (1969): Finite bandwidth, finite amplitude convection. J. Fluid. Mech. **38**, 279

Newell, A. C. (1974): Envelope equations. Lectures in Appl. Math. **15**, 157

Nicolis, G., Prigogine, I. (1977): *Self-Organization in Nonequilibrium Systems – From Dissipative Structures to Order through Fluctuations* (Wiley, New York)

Ortoleva, P., Ross, J. (1973): Phase waves in oscillating chemical reactions. J. Chem. Phys. **58**, 5673

Ortoleva, P., Ross, J. (1974): On a variety of wave phenomena in chemical reactions. J. Chem. Phys. **60**, 5090

Ortoleva, P., Ross, J. (1975): Theory of propagation of discontinuities in kinetic systems with multiple time scales: Fronts, front multiplicity, and pulses. J. Chem. Phys. **63**, 3398

Pavlidis, T. (1973): *Biological Oscillators – Their Mathematical Analysis* (Academic, New York)

Pomeau, Y., Roux, J. C., Rossi, A., Bachelart, S., Vidal, C. (1981): Intermittent behavior in the Belousov-Zhabotinsky reaction. J. Phys. (Paris) Lett **42**, L-271

Rashevsky, N. (1940): An approach to the mathematical biophysics of biological self-regulation and of cell polarity. Bull. Math. Biophys. **2**, 15

Rinzel, J., Keller, J. B. (1973): Traveling wave solutions of a nerve conduction equation. Biophys. J. **13**, 1313

Rinzel, J. (1975): Neutrally stable traveling wave solutions of nerve conduction equations. J. Math. Biol. **2**, 205

Rössler, O. E. (1976): Chemical turbulence: Chaos in a simple reaction-diffusion system. Z. Naturforsch. **31a**, 1168

Rössler, O. E. (1977): Chemical turbulence – A synopsis. In *Synergetics – A Workshop*, ed. by H. Haken, Springer Ser. Syn., Vol. 2 (Springer, Berlin, Heidelberg, New York) p. 174

Rössler, O. E., Wegmann, E. (1978): Chaos in Zhabotinskii reaction. Nature **271**, 89

Rössler, O. E., Kahlert, C. (1979): Winfree meandering in a 2-dimensional 2-variable excitable medium. Z. Naturforsch. **34a**, 565

Roux, J. C., Rossi, A., Bachelart, S., Vidal, C. (1981): Experimental observations of complex dynamical behavior during a chemical reaction. Physica **2D**, 395

Satsuma, J. (1981): Exact solutions of a nonlinear diffusion equation. J. Phys. Soc. Japan **50**, 1423

Schmitz, R. A., Graziani, K. R., Hudson, J. L. (1977): Experimental evidence of chaotic states in the Belousov-Zhabotinskii reaction. J. Chem. Phys. **67**, 3040

Segel, L. A., Jackson, J. L. (1972): Dissipative structure: An explanation and an ecological example. J. Theor. Biol. **37**, 545

Simoyi, R. H., Wolf, A., Swinney, H. L. (1982): One-dimensional dynamics in a multicomponent chemical reaction. Phys. Rev. Lett. **49**, 245

Sivashinsky, G. I. (1977): Nonlinear analysis of hydrodynamic instability in laminar flames – I. Derivation of basic equations. Acta Astronautica **4**, 1177

Sivashinsky, G. I. (1979): On self-turbulization of a laminar flame. Acta Astronautica **6**, 569

Stewartson, K., Stuart, J. T. (1971): A non-linear instability theory for a wave system in plane Poiseuille flow. J. Fluid Mech. **48**, 529

Stratonovich, R. L. (1967): *Topics in the Theory of Random Noise* (Gordon and Breach, New York)

Stuart, J. T. (1960): On the nonlinear mechanics of wave disturbances in stable and unstable parallel flows. Part I: The basic behavior in plane Poiseuille flow. J. Fluid Mech. **9**, 353

Suzuki, R. (1976): Electrochemical neuron model. Adv. Biophys. **9**, 115

Taniuti, T., Wei, C. C. (1968): Reductive perturbation method in nonlinear wave propagation. I. J. Phys. Soc. Japan **24**, 941

Taniuti, T. (1974): Reductive perturbation method and far fields of wave equations. Prog. Theor. Phys. Suppl. **55**, 1

Tomita, K., Tomita, H. (1974): Irreversible circulation of fluctuation. Prog. Theor. Phys. **51**, 1731
Tomita, K., Tsuda, I. (1980): Towards the interpretation of Hudson's experiment of the Belousov-Zhabotinsky reaction. Prog. Theor. Phys. **64**, 1138
Turing, A. M. (1952): The chemical basis of morphogenesis. Phil. Trans. Roy. Soc. London **B237**, 37
Turner, J. S., Roux, J. C., McCormick, W. D., Swinney, H. L. (1981): Alternating periodic and chaotic regimes in a chemical reaction – Experiment and theory. Phys. Lett. **85A**, 9
Tyson, J. J. (1976): *The Belousov-Zhabotinskii Reaction*, Lecture Notes Biomath., Vol. 10 (Springer, Berlin, Heidelberg, New York)
Vidal, C., Roux, J. C., Bachelart, S., Rossi, A. (1980): Experimental study of the transition to turbulence in the Belousov-Zhabotinsky reaction. In *Nonlinear Dynamics*, Ann. NY Acad. Sci., Vol. 357, ed. by R. H. G. Helleman (NY Acad. Sci., New York) p. 377
Wiener, N. (1965): *Nonlinear Problems in Random Theory*, 2nd ed. (M. I. T. Press, Boston)
Winfree, A. T. (1967): Biological rhythms and the behavior of populations of coupled oscillators. J. Theor. Biol. **16**, 15
Winfree, A. T. (1972): Spiral waves of chemical activity. Science **175**, 634
Winfree, A. T. (1974a): Rotating chemical reactions. Sci. Am. **230**, 82
Winfree, A. T. (1974b): Two kinds of waves in an oscillating chemical solution. Farad. Symp. Chem. Soc. **9**, 38
Winfree, A. T. (1978): Stably rotating patterns of reaction and diffusion. In *Theoretical Chemistry*, Vol. 4, ed. by H. Eyring, D. Henderson (Academic, New York) p. 1
Winfree, A. T. (1980): *The Geometry of Biological Time*, Biomath. Vol. 8 (Springer, New York)
Wunderlin, A., Haken, H. (1975): Scaling theory of nonequilibrium systems. Z. Phys. **B21**, 393
Yakhot, V. (1981): Large-scale properties of unstable systems governed by the Kuramoto-Sivashinsky equation. Phys. Rev. **A24**, 642
Yamada, T., Kuramoto, Y. (1976a): Spiral waves in a nonlinear dissipative system. Prog. Theor. Phys. **55**, 2035
Yamada, T., Kuramoto, Y. (1976b): A reduced model showing chemical turbulence. Prog. Theor. Phys. **56**, 681
Yamaguchi, Y., Kometani, K., Shimizu, H. (1981): Self-synchronization of nonlinear oscillations in the presence of fluctuations. J. Stat. Phys. **26**, 719
Yamazaki, H., Oono, Y., Hirakawa, K. (1978): Experimental study on chemical turbulence. J. Phys. Soc. Japan **44**, 335
Yamazaki, H., Oono, Y., Hirakawa, K. (1979): Experimental study of chemical turbulence. II. J. Phys. Soc. Japan **46**, 721
Yorke, J. A., Yorke, E. D. (1979): Metastable chaos: The transition to sustained chaotic behavior in the Lorenz model. J. Stat. Phys. **21**, 263
Zaikin, A. N., Zhabotinsky, A. M. (1970): Concentration wave propagation in two-dimensional liquid-phase self-oscillating systems. Nature **225**, 535
Zhabotinsky, A. M. (1974): *Spontaneously Oscillating Concentrations* (Science Publishers, Moscow) (in Russian)

Subject Index

A CATALOG OF SELECTED
DOVER BOOKS
IN SCIENCE AND MATHEMATICS

Chemistry

THE SCEPTICAL CHYMIST: THE CLASSIC 1661 TEXT, Robert Boyle. Boyle defines the term "element," asserting that all natural phenomena can be explained by the motion and organization of primary particles. 1911 ed. viii+232pp. 5⅜ x 8½.
0-486-42825-7

RADIOACTIVE SUBSTANCES, Marie Curie. Here is the celebrated scientist's doctoral thesis, the prelude to her receipt of the 1903 Nobel Prize. Curie discusses establishing atomic character of radioactivity found in compounds of uranium and thorium; extraction from pitchblende of polonium and radium; isolation of pure radium chloride; determination of atomic weight of radium; plus electric, photographic, luminous, heat, color effects of radioactivity. ii+94pp. 5⅜ x 8½. 0-486-42550-9

CHEMICAL MAGIC, Leonard A. Ford. Second Edition, Revised by E. Winston Grundmeier. Over 100 unusual stunts demonstrating cold fire, dust explosions, much more. Text explains scientific principles and stresses safety precautions. 128pp. 5⅜ x 8½. 0-486-67628-5

THE DEVELOPMENT OF MODERN CHEMISTRY, Aaron J. Ihde. Authoritative history of chemistry from ancient Greek theory to 20th-century innovation. Covers major chemists and their discoveries. 209 illustrations. 14 tables. Bibliographies. Indices. Appendices. 851pp. 5⅜ x 8½. 0-486-64235-6

CATALYSIS IN CHEMISTRY AND ENZYMOLOGY, William P. Jencks. Exceptionally clear coverage of mechanisms for catalysis, forces in aqueous solution, carbonyl- and acyl-group reactions, practical kinetics, more. 864pp. 5⅜ x 8½.
0-486-65460-5

ELEMENTS OF CHEMISTRY, Antoine Lavoisier. Monumental classic by founder of modern chemistry in remarkable reprint of rare 1790 Kerr translation. A must for every student of chemistry or the history of science. 539pp. 5⅜ x 8½. 0-486-64624-6

THE HISTORICAL BACKGROUND OF CHEMISTRY, Henry M. Leicester. Evolution of ideas, not individual biography. Concentrates on formulation of a coherent set of chemical laws. 260pp. 5⅜ x 8½. 0-486-61053-5

A SHORT HISTORY OF CHEMISTRY, J. R. Partington. Classic exposition explores origins of chemistry, alchemy, early medical chemistry, nature of atmosphere, theory of valency, laws and structure of atomic theory, much more. 428pp. 5⅜ x 8½. (Available in U.S. only.) 0-486-65977-1

GENERAL CHEMISTRY, Linus Pauling. Revised 3rd edition of classic first-year text by Nobel laureate. Atomic and molecular structure, quantum mechanics, statistical mechanics, thermodynamics correlated with descriptive chemistry. Problems. 992pp. 5⅜ x 8½. 0-486-65622-5

FROM ALCHEMY TO CHEMISTRY, John Read. Broad, humanistic treatment focuses on great figures of chemistry and ideas that revolutionized the science. 50 illustrations. 240pp. 5⅜ x 8½. 0-486-28690-8

Engineering

DE RE METALLICA, Georgius Agricola. The famous Hoover translation of greatest treatise on technological chemistry, engineering, geology, mining of early modern times (1556). All 289 original woodcuts. 638pp. 6¾ x 11. 0-486-60006-8

FUNDAMENTALS OF ASTRODYNAMICS, Roger Bate et al. Modern approach developed by U.S. Air Force Academy. Designed as a first course. Problems, exercises. Numerous illustrations. 455pp. 5⅜ x 8½. 0-486-60061-0

DYNAMICS OF FLUIDS IN POROUS MEDIA, Jacob Bear. For advanced students of ground water hydrology, soil mechanics and physics, drainage and irrigation engineering and more. 335 illustrations. Exercises, with answers. 784pp. 6⅛ x 9¼. 0-486-65675-6

THEORY OF VISCOELASTICITY (Second Edition), Richard M. Christensen. Complete consistent description of the linear theory of the viscoelastic behavior of materials. Problem-solving techniques discussed. 1982 edition. 29 figures. xiv+364pp. 6⅛ x 9¼. 0-486-42880-X

MECHANICS, J. P. Den Hartog. A classic introductory text or refresher. Hundreds of applications and design problems illuminate fundamentals of trusses, loaded beams and cables, etc. 334 answered problems. 462pp. 5⅜ x 8½. 0-486-60754-2

MECHANICAL VIBRATIONS, J. P. Den Hartog. Classic textbook offers lucid explanations and illustrative models, applying theories of vibrations to a variety of practical industrial engineering problems. Numerous figures. 233 problems, solutions. Appendix. Index. Preface. 436pp. 5⅜ x 8½. 0-486-64785-4

STRENGTH OF MATERIALS, J. P. Den Hartog. Full, clear treatment of basic material (tension, torsion, bending, etc.) plus advanced material on engineering methods, applications. 350 answered problems. 323pp. 5⅜ x 8½. 0-486-60755-0

A HISTORY OF MECHANICS, René Dugas. Monumental study of mechanical principles from antiquity to quantum mechanics. Contributions of ancient Greeks, Galileo, Leonardo, Kepler, Lagrange, many others. 671pp. 5⅜ x 8½. 0-486-65632-2

STABILITY THEORY AND ITS APPLICATIONS TO STRUCTURAL MECHANICS, Clive L. Dym. Self-contained text focuses on Koiter postbuckling analyses, with mathematical notions of stability of motion. Basing minimum energy principles for static stability upon dynamic concepts of stability of motion, it develops asymptotic buckling and postbuckling analyses from potential energy considerations, with applications to columns, plates, and arches. 1974 ed. 208pp. 5⅜ x 8½. 0-486-42541-X

METAL FATIGUE, N. E. Frost, K. J. Marsh, and L. P. Pook. Definitive, clearly written, and well-illustrated volume addresses all aspects of the subject, from the historical development of understanding metal fatigue to vital concepts of the cyclic stress that causes a crack to grow. Includes 7 appendixes. 544pp. 5⅜ x 8½. 0-486-40927-9

Mathematics

FUNCTIONAL ANALYSIS (Second Corrected Edition), George Bachman and Lawrence Narici. Excellent treatment of subject geared toward students with background in linear algebra, advanced calculus, physics and engineering. Text covers introduction to inner-product spaces, normed, metric spaces, and topological spaces; complete orthonormal sets, the Hahn-Banach Theorem and its consequences, and many other related subjects. 1966 ed. 544pp. 6⅛ x 9¼. 0-486-40251-7

ASYMPTOTIC EXPANSIONS OF INTEGRALS, Norman Bleistein & Richard A. Handelsman. Best introduction to important field with applications in a variety of scientific disciplines. New preface. Problems. Diagrams. Tables. Bibliography. Index. 448pp. 5⅜ x 8½. 0-486-65082-0

VECTOR AND TENSOR ANALYSIS WITH APPLICATIONS, A. I. Borisenko and I. E. Tarapov. Concise introduction. Worked-out problems, solutions, exercises. 257pp. 5⅝ x 8¼. 0-486-63833-2

AN INTRODUCTION TO ORDINARY DIFFERENTIAL EQUATIONS, Earl A. Coddington. A thorough and systematic first course in elementary differential equations for undergraduates in mathematics and science, with many exercises and problems (with answers). Index. 304pp. 5⅜ x 8½. 0-486-65942-9

FOURIER SERIES AND ORTHOGONAL FUNCTIONS, Harry F. Davis. An incisive text combining theory and practical example to introduce Fourier series, orthogonal functions and applications of the Fourier method to boundary-value problems. 570 exercises. Answers and notes. 416pp. 5⅜ x 8½. 0-486-65973-9

COMPUTABILITY AND UNSOLVABILITY, Martin Davis. Classic graduate-level introduction to theory of computability, usually referred to as theory of recurrent functions. New preface and appendix. 288pp. 5⅜ x 8½. 0-486-61471-9

ASYMPTOTIC METHODS IN ANALYSIS, N. G. de Bruijn. An inexpensive, comprehensive guide to asymptotic methods–the pioneering work that teaches by explaining worked examples in detail. Index. 224pp. 5⅜ x 8½ 0-486-64221-6

APPLIED COMPLEX VARIABLES, John W. Dettman. Step-by-step coverage of fundamentals of analytic function theory–plus lucid exposition of five important applications: Potential Theory; Ordinary Differential Equations; Fourier Transforms; Laplace Transforms; Asymptotic Expansions. 66 figures. Exercises at chapter ends. 512pp. 5⅜ x 8½. 0-486-64670-X

INTRODUCTION TO LINEAR ALGEBRA AND DIFFERENTIAL EQUATIONS, John W. Dettman. Excellent text covers complex numbers, determinants, orthonormal bases, Laplace transforms, much more. Exercises with solutions. Undergraduate level. 416pp. 5⅜ x 8½. 0-486-65191-6

RIEMANN'S ZETA FUNCTION, H. M. Edwards. Superb, high-level study of landmark 1859 publication entitled "On the Number of Primes Less Than a Given Magnitude" traces developments in mathematical theory that it inspired. xiv+315pp. 5⅜ x 8½. 0-486-41740-9

Math–Decision Theory, Statistics, Probability

Physics

OPTICAL RESONANCE AND TWO-LEVEL ATOMS, L. Allen and J. H. Eberly. Clear, comprehensive introduction to basic principles behind all quantum optical resonance phenomena. 53 illustrations. Preface. Index. 256pp. 5⅜ x 8½. 0-486-65533-4

QUANTUM THEORY, David Bohm. This advanced undergraduate-level text presents the quantum theory in terms of qualitative and imaginative concepts, followed by specific applications worked out in mathematical detail. Preface. Index. 655pp. 5⅜ x 8½. 0-486-65969-0

ATOMIC PHYSICS (8th EDITION), Max Born. Nobel laureate's lucid treatment of kinetic theory of gases, elementary particles, nuclear atom, wave-corpuscles, atomic structure and spectral lines, much more. Over 40 appendices, bibliography. 495pp. 5⅜ x 8½. 0-486-65984-4

A SOPHISTICATE'S PRIMER OF RELATIVITY, P. W. Bridgman. Geared toward readers already acquainted with special relativity, this book transcends the view of theory as a working tool to answer natural questions: What is a frame of reference? What is a "law of nature"? What is the role of the "observer"? Extensive treatment, written in terms accessible to those without a scientific background. 1983 ed. xlviii+172pp. 5⅜ x 8½. 0-486-42549-5

AN INTRODUCTION TO HAMILTONIAN OPTICS, H. A. Buchdahl. Detailed account of the Hamiltonian treatment of aberration theory in geometrical optics. Many classes of optical systems defined in terms of the symmetries they possess. Problems with detailed solutions. 1970 edition. xv + 360pp. 5⅜ x 8½. 0-486-67597-1

PRIMER OF QUANTUM MECHANICS, Marvin Chester. Introductory text examines the classical quantum bead on a track: its state and representations; operator eigenvalues; harmonic oscillator and bound bead in a symmetric force field; and bead in a spherical shell. Other topics include spin, matrices, and the structure of quantum mechanics; the simplest atom; indistinguishable particles; and stationary-state perturbation theory. 1992 ed. xiv+314pp. 6⅛ x 9¼. 0-486-42878-8

LECTURES ON QUANTUM MECHANICS, Paul A. M. Dirac. Four concise, brilliant lectures on mathematical methods in quantum mechanics from Nobel Prize-winning quantum pioneer build on idea of visualizing quantum theory through the use of classical mechanics. 96pp. 5⅜ x 8½. 0-486-41713-1

THIRTY YEARS THAT SHOOK PHYSICS: THE STORY OF QUANTUM THEORY, George Gamow. Lucid, accessible introduction to influential theory of energy and matter. Careful explanations of Dirac's anti-particles, Bohr's model of the atom, much more. 12 plates. Numerous drawings. 240pp. 5⅜ x 8½. 0-486-24895-X

ELECTRONIC STRUCTURE AND THE PROPERTIES OF SOLIDS: THE PHYSICS OF THE CHEMICAL BOND, Walter A. Harrison. Innovative text offers basic understanding of the electronic structure of covalent and ionic solids, simple metals, transition metals and their compounds. Problems. 1980 edition. 582pp. 6⅛ x 9¼. 0-486-66021-4

A TREATISE ON ELECTRICITY AND MAGNETISM, James Clerk Maxwell. Important foundation work of modern physics. Brings to final form Maxwell's theory of electromagnetism and rigorously derives his general equations of field theory. 1,084pp. 5⅜ x 8½. Two-vol. set. Vol. I: 0-486-60636-8 Vol. II: 0-486-60637-6

QUANTUM MECHANICS: PRINCIPLES AND FORMALISM, Roy McWeeny. Graduate student-oriented volume develops subject as fundamental discipline, opening with review of origins of Schrödinger's equations and vector spaces. Focusing on main principles of quantum mechanics and their immediate consequences, it concludes with final generalizations covering alternative "languages" or representations. 1972 ed. 15 figures. xi+155pp. 5⅜ x 8½. 0-486-42829-X

INTRODUCTION TO QUANTUM MECHANICS With Applications to Chemistry, Linus Pauling & E. Bright Wilson, Jr. Classic undergraduate text by Nobel Prize winner applies quantum mechanics to chemical and physical problems. Numerous tables and figures enhance the text. Chapter bibliographies. Appendices. Index. 468pp. 5⅜ x 8½. 0-486-64871-0

METHODS OF THERMODYNAMICS, Howard Reiss. Outstanding text focuses on physical technique of thermodynamics, typical problem areas of understanding, and significance and use of thermodynamic potential. 1965 edition. 238pp. 5⅜ x 8½. 0-486-69445-3

THE ELECTROMAGNETIC FIELD, Albert Shadowitz. Comprehensive undergraduate text covers basics of electric and magnetic fields, builds up to electromagnetic theory. Also related topics, including relativity. Over 900 problems. 768pp. 5⅝ x 8¼. 0-486-65660-8

GREAT EXPERIMENTS IN PHYSICS: FIRSTHAND ACCOUNTS FROM GALILEO TO EINSTEIN, Morris H. Shamos (ed.). 25 crucial discoveries: Newton's laws of motion, Chadwick's study of the neutron, Hertz on electromagnetic waves, more. Original accounts clearly annotated. 370pp. 5⅜ x 8½. 0-486-25346-5

EINSTEIN'S LEGACY, Julian Schwinger. A Nobel Laureate relates fascinating story of Einstein and development of relativity theory in well-illustrated, nontechnical volume. Subjects include meaning of time, paradoxes of space travel, gravity and its effect on light, non-Euclidean geometry and curving of space-time, impact of radio astronomy and space-age discoveries, and more. 189 b/w illustrations. xiv+250pp. 8⅜ x 9¼. 0-486-41974-6

STATISTICAL PHYSICS, Gregory H. Wannier. Classic text combines thermodynamics, statistical mechanics and kinetic theory in one unified presentation of thermal physics. Problems with solutions. Bibliography. 532pp. 5⅜ x 8½. 0-486-65401-X